Subjunctive Aesthetics

CRITICAL MEXICAN STUDIES

Critical Mexican Studies
Series editor: Ignacio M. Sánchez Prado

Critical Mexican Studies is the first English-language, humanities-based, theoretically focused academic series devoted to the study of Mexico. The series is a space for innovative works in the humanities that focus on theoretical analysis, transdisciplinary interventions, and original conceptual framing.

Other titles in the series:

The Restless Dead: Necrowriting and Disappropriation, by Cristina Rivera Garza

History and Modern Media: A Personal Journey, by John Mraz

Toxic Loves, Impossible Futures: Feminist Living as Resistance, by Irmgard Emmelhainz

Drug Cartels Do Not Exist: Narcotrafficking in US and Mexican Culture, by Oswaldo Zavala

Unlawful Violence: Mexican Law and Cultural Production, by Rebecca Janzen

The Mexican Transpacific: Nikkei Writing, Visual Arts, and Performance, by Ignacio López-Calvo

Monstrous Politics: Geography, Rights, and the Urban Revolution in Mexico City, by Ben Gerlofs

Robo Sacer: Necroliberalism and Cyborg Resistance in Mexican and Chicanx Dystopias by David Dalton

Mexico, Interrupted: Labor, Idleness, and the Economic Imaginary of Independence by Sergio Gutiérrez Negrón

Serial Mexico: Storytelling across Media, from Nationhood to Now by Amy E. Wright

Sonic Strategies: Performing Mexico's War on Drugs, Mourning, and Feminicide by Christina Baker

Subjunctive Aesthetics
Mexican Cultural Production
in the Era of Climate Change

Carolyn Fornoff

Vanderbilt University Press
Nashville, Tennessee

Copyright 2024 Vanderbilt University Press
All rights reserved
First printing 2024

This book will be made open access within three years of publication thanks to Path to Open, a program developed in partnership between JSTOR, the American Council of Learned Societies (ACLS), University of Michigan Press, and the University of North Carolina Press to bring about equitable access and impact for the entire scholarly community, including authors, researchers, libraries, and university presses around the world. Learn more at https://about.jstor.org/path-to-open/.

This work was supported in part by the Hull Memorial Publication Fund of Cornell University.

Library of Congress Cataloging-in-Publication Data
Names: Fornoff, Carolyn, author.
Title: Subjunctive aesthetics : Mexican cultural production in the era of climate change / Carolyn Fornoff.
Description: Nashville, Tennessee : Vanderbilt University Press, [2024] | Series: Critical Mexican studies | Includes bibliographical references and index.
Identifiers: LCCN 2023036850 (print) | LCCN 2023036851 (ebook) | ISBN 9780826506177 (paperback) | ISBN 9780826506184 (hardcover) | ISBN 9780826506191 (epub) | ISBN 9780826506207 (pdf)
Subjects: LCSH: Environment (Aesthetics) | Aesthetics, Mexican. | Aesthetics--Political aspects--Mexico.
Classification: LCC BH221.M6 F67 2024 (print) | LCC BH221.M6 (ebook) | DDC 111/.850972--dc23/eng/20231027
LC record available at https://lccn.loc.gov/2023036850
LC ebook record available at https://lccn.loc.gov/2023036851

For past, present, and future land defenders.

Contents

Acknowledgments ix

   INTRODUCTION   1

1 Environmental Rewriting   25

2 Land Defense and Counterfactual Mourning   57

3 Extinction Poetics   91

4 The Rural Resilience Film   117

5 Greening Mexican Cinema   149

   CONCLUSION   175

Notes 183
Bibliography 225
Index 251

Acknowledgments

I have been fortunate to have many mentors whose support made this book possible. I met Beatriz González-Stephan as an undergraduate student at Rice University; I would not have pursued a doctorate without her encouragement. We have kept in touch, and Beatriz has been a constant cheerleader of my work. I am grateful to Román de la Campa, whose ability to map theory and literature continues to inspire. Bethany Wiggin brought me on as the first graduate student coordinator of the Penn Program in Environmental Humanities, introducing me to interdisciplinary work and site-based practice. Finally, I am grateful to Ignacio Sánchez Prado. Over many a mezcal, Nacho imparted invaluable advice. It is an honor to publish this book in his series.

This book grew out of conversations with countless friends and colleagues. I am grateful to Jorge Téllez, Mark Anderson, Gisela Heffes, and Arturo Arias for their support of my work. To Lindsey Reuben, Kristen Turpin, Steve Dolph, Isabel Díaz, Ana Almeyda Cohen, Ruthie Meadows, and Patricia Kim, thank you for your friendship during grad school. For your solidarity throughout the pandemic in Champaign-Urbana, thank you to Jess Hammie, Patrick Earl Hammie, Jamie Jones, Eric Calderwood, Eduardo Ledesma, Jill Jegerski, Xiomara Cervantes-Gómez, Pollyanna Rhee, and Rebecca Oh. I have found in Mexicanist studies a generous sense of community and support. Thank you to Adriana Pérez Limón, Olivia Cosentino, Iván Eusebio Aguirre Darancou, Fran Dennstedt, Sophie Esch, Rafael Acosta, Jorge Quintana-Navarrete, Roberto Cruz Arzabal, Sergio Gutiérrez Negrón, Mónica García-Blizzard, Miguel Valerio, Pavel Andrade, Brian Price, Brian Gollnick, Adela Pineda Franco, Cheyla Samuelson, and many others. My deepest appreciation to Sara Poot-Herrera, Jacobo Sefamí, and Viviane Mahieux for nurturing the hemispheric Mexicanist community of scholars.

I could not have completed this manuscript without the time to write granted to me by the Humanities Teaching Release at the University of Illinois at Urbana-Champaign (UIUC) and the UIUC Humanities Research Institute (HRI) Fellowship. The support of the UIUC Campus Research Board and the HRI Summer Faculty Research Fellowship were also essential. Thank you to Antoinette Burton and Nancy Castro for believing in my work. I am grateful to my colleagues in the Department of Spanish and Portuguese at UIUC for their support as I drafted this book: Eduardo Ledesma, Xiomara Cervantes-Gómez, Ana Torres-Cacoullos, Javier Irigoyen-García, Pilar Martínez-Quiroga, Elena Delgado, John Karam, Mariselle Meléndez, and Dara Goldman—may her memory be for a blessing. Thank you to my new colleagues in Spanish at Cornell for the warm welcome: Edmundo Paz-Soldán, Simone Pinet, Debra Castillo, Patricia Keller, Julia Chang, Liliana Colanzi, Irina Troconis, and Vanessa Gubbins.

The feedback that I received on this manuscript made it stronger. My feminist ecocriticism writing group generously read and commented on several chapters: Ashley Brock, Martina Broner, Carolina Sá Carvalho, Victoria Saramago, and Amanda Smith. The UIUC environmental humanities group helped me clarify my thoughts and know when to leave things be: Jamie Jones, Rebecca Oh, Pollyanna Rhee, John Levi Barnard, Clara Bosak-Schroeder, and Bob Morrissey. The Rice animal studies writing group provided insight on Chapter 3: Sophie Esch, Kelly McKisson, and Bren Ram. The HRI "Symptoms of Crisis" fellows helped me work through an iteration of Chapter 2. Olivia Cosentino generously read a messy draft of Chapter 4.

Two regular writing groups kept me on track as I drafted and revised this book during the COVID-19 pandemic. Somehow Katie Sobering, Laura Smithers, and Sarah Stanlick did not get tired of my corny jokes on Zoom. My WhatsApp check-in group encouraged me to keep at it: Rebeca Hey-Colón, Emily Hind, Rebecca Janzen, Carmen Serrano, and John Waldron.

The ability to share this work with others has been essential to its evolution. Thank you to Iñaki Prádanos-García at Miami University, Kerstin Oloff at Durham University, Andrew Rajka at the University of South Carolina, Emily Hind at the University of Florida, Michelle Bastian at University of Edinburgh, Pavel Andrade at the Mexican Studies Research Collective, Aníbal González at Yale, and Stephen Tobin at UCLA for inviting me to speak about my research and for your audience's feedback.

Collaborating with Vanderbilt University Press has been a wonderful experience. I am grateful to Zack Gresham for shepherding me through the process, and to Nacho for including me in the Critical Mexican Studies series. The comments made by the anonymous peer reviewers were spot on; thank you for your rigor. Arkaitz Ibarretxe Diego helped pore over my works cited. Diana Rico's keen eye was essential during the final stages of preparing the manuscript for publication. Thank you to Andrew Ascherl for your skillful indexing.

Part of Chapter 3 is derived from "Planetary Poetics of Extinction in Contemporary Mexican Poetry" in *Mexican Literature as World Literature*, edited by Ignacio M. Sánchez Prado, Bloomsbury Press, 2021. An abridged version of Chapter 5 is published as "Greening Mexican Cinema" in *Ecocinema Theory and Practice 2*, edited by Stephen Rust, Salma Monani, and Seán Cubitt, Routledge, 2022. I am grateful to the authors and artists who allowed me to republish their work in this book. Thank you to Marcela Armas, Verónica Gerber Bicecci, Mikeas Sánchez and Wendy Call, Roberto Vega and Rosario Martinez of Lapiztola, Naomi Rincón Gallardo, Diego Torres and Rodrigo Soto of Cine Móvil ToTo, and the Archivo José Juan Tablada.

The roots of a project go back much further of course. My mother used to line edit my papers in high school, instilling in me the pleasure of a well-crafted sentence. My parents encouraged me to study abroad for a year in Mendoza at the age of fifteen; an experience with formative impact, to say the least. Suzi, I am delighted to live closer to you now to see Harley grow up. Most of all, thank you to Jon for accompanying me on this journey. From the tears shed over my first grad school papers, to two leg surgeries, and three moves for my career, I owe you big time. Now I can finally stop talking about "the book." Or at least this one.

Introduction

In the poem "Piel territorio" (Skin territory), Zapotec filmmaker Luna Marán writes, "la tierra es más allá de nosotros, / pero nosotros no somos sin ella" (the land exceeds us / but we do not exist without it).[1] Unlike Western conceptualizations of land as property or as resource, Marán elaborates that for the Zapotec, territory is a skin: a protective covering and porous interface that is inextricable from the self. Marán writes "Piel territorio" in the context of Indigenous dispossession in Mexico by developmental megaprojects and extractive industries like mining, to make the case that dispossession is akin to ripping away a community's skin or tongue, a violence that results in the erosion of life and culture.

Marán's poem uses the subjunctive mood to describe the continual process of becoming in relation to the land: "la tierra es eso que hace que *seamos*" (the land is that which *makes us be*).[2] This phrasing situates land as the subject that acts upon the "we" in the subordinate subjunctive clause, and articulates this "we" as indeterminate, always in a state of production in relation with the land. Here, the subjunctive mood encodes possibility and dependency, the contingent way that individuals, communities, and environments come into contact and transform each other in the process. "La tierra es eso que hace que seamos" concatenates land with the possibility of Zapotec community. Extrapolated to the context of planetary climate change, the verse models a subjunctive orientation to territory as a determining factor around which life is variably assembled.

With this book, I set out to assess the wide array of literary, visual, and filmic arts produced in twenty-first century Mexico in response to socio-environmental crisis. I wanted to understand the narrative and aesthetic strategies that artists, filmmakers, and writers use to narrate ecological catastrophe, and specifically situations that foreclose the possibility of future life like dispossession and extinction. What I found, as in "Piel territorio," is the recurrence of subjunctive modes of artistic expression that contest the definitiveness of foreclosure and instead mobilize desire, emotion, and the imagination to invoke the potential of a postextractivist future. As a political project, postextractivism aims to unravel extractivist paradigms entrenched since the colonial period that link modernity with the exploitation and export of nonhuman nature, and that subsume life to profit. In a context in which extractivism is continually asserted across the political spectrum as the only possible path to collective well-being, the subjunctive ability of art to imagine the world otherwise becomes a key means by which to express alternative formulations of how the relationship with territory *should be* or *could be*.

This book theorizes the subjunctive as the defining mood of contemporary Mexican cultural responses to environmental crisis, from extractivist dispossession to climate change. In contrast with the indicative, the grammar of *what is*, the subjunctive mood is the realm of the potential and the uncertain. The Real Academia Española defines the subjunctive as a grammatical "modo con que se marca lo expresado por el predicado como información virtual, inespecífica, no verificada, o no experimentada" (mode that marks what has been expressed by the predicate as virtual, unspecific, unverified, or unrealized information).[3] Semantically motivated and lexically mandated, in the Spanish language the subjunctive has numerous and even contradictory uses, so much so that textbooks targeting second-language learners urge students to "concentrate on learning *when* to use the subjunctive rather than on asking *why* one uses it or what it 'means.'"[4] The subjunctive is perhaps the most difficult grammatical concept for second-language learners of Spanish to master. This is because in modern English, even though the subjunctive exists as a grammatical mood and is signaled by auxiliaries such as *might*, *could*, *should*, or *would*, it does not recognizably impact verb conjugation. This decreases English speakers' awareness of its presence and function, and it heightens the difficulty of mastery in Spanish, a language in which the subjunctive mood does require conjugation. This inexperience is compounded by the disorienting constellation of uses for the subjunctive in Spanish, an array that is more varied than in any other Romance language and that evades attempts to corral it under an intuitive set of rules.

While sweeping definitions risk obscuring the nuance that is characteristic of linguistic systems, grammatical moods can be roughly parsed into two categories, *realis* and *irrealis*. Realis, associated with the indicative, is the mood of evidence and truth. Through the indicative, the speaker confidently affirms the facticity and definitiveness of their claims. By contrast, irrealis moods, including the subjunctive, indicate no such commitment to a singular or fixed truth.[5] The subjunctive encodes the speaker's uncertainty or opinionating about the state of things. It is used to respond to reality in a way that expresses its potential to change—or the speaker's desire for it to change—whether that be the contingency of interacting lifeforms registered by Marán's poem, or its thematic treatment of the yearning for a world without dispossession. Translated to the realm of aesthetics, subjunctive aesthetics are less invested in accuracy and reportage, and more in the desires and fears that the real provokes, as well as the as-of-yet unrealized possibilities that percolate around the given.

Environmental crisis compels questions in the subjunctive, about how things *might* but *need not be*. What would it take to curb global emissions? And how could the world be organized differently, around life rather than around accumulation by dispossession? Thinking in the subjunctive explodes the fixity of the way things are, suggesting, even against all odds, that the status quo can be transformed. As Kierkegaard put it, "when one begins to study the grammar of the indicative and the subjunctive . . . one becomes conscious that everything depends on how it is thought. The indicative thinks something as actual. . . . The subjunctive thinks something as thinkable."[6] For its ability to make things thinkable, the subjunctive has long motored the arts as a site where reality brushes up against and is tested against ideality, hypothesis, and probability.

Despite the innate alignment of the arts and the subjunctive, the subjunctive turn in contemporary cultural production about environmental crisis is somewhat surprising given the historical primacy of indicative modes of truth-telling in environmentalist art. In the context of environmental crisis, literature and visual art have often been valorized for their evidentiary, mimetic, and didactic force: the way that representation makes the invisible visible or raises public awareness to the facts. Such pedagogical or forensic aesthetics are of course hugely valuable. As Natalia Mendoza argues, they are necessitated by deadly incidents like the forced disappearance of forty-three students from the Ayotzinapa Rural Teacher's College and the toxic waste spill in the Sonora River, the worst mining disaster in Mexican history, both of which took place in 2014.[7] In the face of state complicity and impunity, forensic representational practices take up the urgent task of investigating the truth, reconstructing the events, and providing the public with closure by making sense of the senseless. Forensic aesthetics have productively reframed overlooked disasters, like Yuri Herrera's 2018 book *El incendio de la mina El Bordo* (*A Silent Fury: The El Bordo Mine Fire*), which scrutinizes a fire that took place in a Hidalgo mine in 1920 to problematize its silencing at the time by company and state officials.[8]

Subjunctive Aesthetics is not an argument against realist, forensic, or evidentiary strategies of truth telling but rather a celebration of the complementary power of thinking in the subjunctive and its reactive modes of revision, negation, and hypothesis. Subjunctive aesthetics can be understood as a generative counterpoint to forensic aesthetics. Both respond to situations of material damage and the violent foreclosure of life. But whereas forensic aesthetics sustain an anthropological focus on witnessing and marshal material evidence toward establishing a shared truth, subjunctive aesthetics embrace uncertainty, anxiety, and ambiguity in the service of postextractivist imagining. Thinking in the subjunctive tilts away from narrative certainty and moves toward knowledge-making practices motored by doubt, emotion, and imagination. It is one way of getting at Marisol de la Cadena's argument that we should aim "to displace the knowing *anthropos*, to enable knowing-feeling-thinking without it."[9]

Subjunctive aesthetics invite interpretation and debate. It is an especially helpful strategy in this era of communicative capitalism, which Irmgard Emmelhainz explains is characterized by the uncritical experience of the world through images that are presented as detached from any specific point of view and consumed as "empty sensations or tautological truths about reality."[10] The task of the critic and the artist, according to Emmelhainz, is to push back against the idea that the image provides an unmediated glimpse of reality and instead stress its subjectivity. Like Emmelhainz, I am interested in thinking about what art can do beyond its evidentiary function, beyond making the facts of environmental crisis more incontrovertible. Rather than see art as serving an ancillary, illustrative, or didactic role in proving environmental damage, subjunctive aesthetics make a bid for art's experimental and experiential capacity to generate alternative narratives, values, and grammars of territorial belonging. These works are not necessarily encoded in the subjunctive grammatical mood, but are in sync with its precepts, centering the role of desire in producing the future or mobilizing the "as if" to imagine other ways of relating to territory.

In this sense, subjunctive aesthetics align with what Alejandra Amatto has described as the rise of "descontento realista" (realist discontent) in contemporary Latin American literature, or the forceful reemergence of nonrealist genres like the fantastic, sci-fi, and horror as the means by which to channel generalized discontentment with the real.[11] Speculative genres are indeed motored by what Samuel Delany calls "subjunctivity."[12] Amitav Ghosh in *The Great Derangement* goes even further. He pointedly writes that to reproduce the world as it exists need not be the project of fiction; what fiction ... makes possible is to approach the world in a subjunctive mode, to conceive of it as if it were other than it is: in short, the great, irreplaceable potentiality of fiction is that it makes possible the imagining of possibilities. And to imagine other forms of human existence is exactly the challenge that is posed by the climate crisis: for if there is any one thing that global warming has made perfectly clear it is that to think about the world only as it is amounts to a formula for collective suicide. We need, rather, to envision what it might be.[13]

Ghosh contends that fiction is uniquely able to push readers to aspire and dream in the tradition of *ojalá*, "hopefully," or to hypothesize in the *como si fuera*, or "as if it were." At a time when the future seems to be predetermined, the subjunctive ability of art to make things thinkable enters to make the inevitable more propositional and to concoct other possibilities. Subjunctive thinking thus activates what Anna Lowenhaupt Tsing identifies as the "possibilities of coexistence within environmental disturbance."[14]

While my theorization of subjunctive aesthetics draws on thinkers like Ghosh, I do not use the subjunctive as shorthand for speculative fiction. Speculation is but one modality of subjunctive aesthetics, which I conceptualize more expansively around the coordinates of contingency, supposition, affect, and dependency. While many scholars have commented on the imaginative force of the subjunctive to scramble agreed-upon perceptions of reality and establish ways of seeing and sensing oriented around probability rather than facticity, less has been said about the subjunctive's subordinated positionality. This quality of subordination is foundational to the subjunctive mood, which got its name because of its placement in subjoined or subordinate clauses, as in Marán's phrase "la tierra es eso que hace que seamos." The subjunctive is frequently a postmodifier; it makes its home in the dependent clause, supplementing, supporting, contesting, or complicating what has come before. Its use is determined by the semantic properties of the matrix verb; if this verb asserts factuality, then the subordinate verb is indicative, but if it expresses a wish, doubt, possibility, or emotion, and so on, then the subjunctive is deployed.[15] The subordinated structure of the subjunctive means that it always refers back to an antecedent, something other than itself; it is dictated by contingent external forces. The subjunctive always points away from itself and toward another relation, indexing grammar's structure as an assemblage, a web of coordinates harnessed together.

Extrapolated to the theorization of subjunctive aesthetics in the era of climate change, the dependent and subordinated nature of the subjunctive grammatical structure—its inability to exist on its own, except as a command—speaks to the human dependency on the planet to either feed accumulation or sustain life, to reprise Marán yet again: "la tierra es eso que hace que seamos." The subjunctive mood mirrors the way in which we are compelled to respond to planetary conditions, phenomena put in motion by structures, be they economic or climatic, that supersede us. Our subordination to these determining structures indexes that regardless of how we might react to, negate, or wish the world were different, we cannot reinvent the earth. We can, however, reinvent the way that we relate to it and to one another. Tethered to the reality of the Anthropocene, subjunctive aesthetics simultaneously index our dependency on the past and present but also the possibility that we might reformulate our relationship with the more-than-human planet and conjure into being other futures.

I use the term *subjunctive aesthetics* therefore to describe three intersecting trends in contemporary Mexican literary and visual art about ecological crisis. First is a thematic concern with the foreclosure of the future by extractivist policies and carbon-intensive modernity and the use of subjunctive strategies like doubt, desire, and speculation to contest the definitiveness of that foreclosure. Second, subjunctive aesthetics are less invested in art's evidentiary function, and more in its imaginative potential. And third, like the subordinated structure of the subjunctive grammatical mood, subjunctive aesthetics register a state of subordination and entanglement with external factors from the climatic to the economic. Marán's poem, "Piel territorio," contains all three characteristics of what I call subjunctive aesthetics: dispossession as the foreclosure of Zapotec futurity; poetry as a form that bears witness to dispossession but also suspends resolution, articulating land defense or what Marán describes as "darle de comer a la tierra" (feeding the earth) as an ever unfinished project of futurity: "aquí estamos luchando / para que ellos tengan un lugar donde descansar en paz" (here we are fighting / so that they might have a place to rest in peace); and finally, the dependent entanglement of human life with territory: "la tierra es eso que hace que seamos."[16]

I circumscribe my study to cultural production, broadly construed, produced in Mexico between 2012 and 2022, during the administrations of Enrique Peña Nieto and Andrés Manuel López Obrador (AMLO). During this ten-year period, climate change has become an indisputable issue of domestic concern as historic droughts and rising temperatures have destabilized water and food security. In response, both administrations—representing different political parties, Peña Nieto of the center-right Partido Revolucionario Institucional (PRI, Institutional Revolutionary Party) and AMLO of the left-leaning Movimiento Regeneración Nacional (MORENA,

National Regeneration Movement)—have paid lip service to environmental protections, establishing goals at the 2015 United Nations Climate Change Conference (COP21) in Paris (and reaffirming them in 2021 at COP26 in Glasgow) to reduce greenhouse gas emissions, eliminate deforestation by 2030, promote resilient food systems, and conserve biodiversity.[17] Yet, like other countries around the world, Mexico has seen little progress on these fronts. In fact, the opposite is true: cuts to environmental regulatory bodies have proceeded apace, illustrated by the slashing in half of the Secretariat of the Environment and Natural Resources' (SEMARNAT) budget between 2015 and 2019 and by how both administrations have repeatedly sided with corporate interests over the concerns of citizens.[18] While there are some stark policy divergences between the two administrations when it comes to extractivism—most notably with regard to the state-owned oil company, Petróleos Mexicanos (Pemex), which Peña Nieto aimed to privatize and AMLO aimed to strengthen to renew the country's commitment to energy sufficiency (albeit through investment in new refineries and exploratory fracking, rather than diversification into renewable sources like solar or wind)—taken holistically, we might categorize this period as a moment in which the realities of climate change and environmental degradation have come into heightened contrast with the continuation of business as usual.

The paradoxical coincidence of the accelerated onset of symptoms of climate change and the renewed commitment to extractivist development in Mexico has prompted an outpouring of fine art, literature, performance, new media, and film that foregrounds the ethical imperative to rethink the relationship with territory, just as it has in the Southern Cone, as recently studied by Paula Serafini, and the Andes, as studied by Macarena Gómez-Barris.[19] The artists discussed in this book embrace the uncertainty of this moment as an opportunity to reevaluate entrenched models of development and imagine alternative alignments with the environment as territory. My use of the term *territory* is informed by Indigenous formulations of land defense as "la defensa del territorio." This approach is more conceptually ambitious than that of environmentalism, which typically brackets questions of human and environmental well-being. By contrast, as we can see from Marán's poem, Zapotec practices of land defense formulate the two as inextricable. Territory is a specific material space that is produced and ascribed meaning through its relational use and that, in turn, produces those who live in relation to it.[20] This means that it is impossible to think about territory without subjects or subjects without territory; the two are coconstitutive and cosustaining, just as skin is to the body and the self. Territorial defense, then, is the defense of life and of a project in common.

As the first book-length study to explore the question of climate change and extractivism in relation to Mexican literature, film, and art, this book invites scholars of Mexican culture to view environmental concerns as at the crux of Mexico's present and imagined future. *Subjunctive Aesthetics* demonstrates that territory is inextricable from key issues in Mexican cultural studies, including coloniality, violence, Indigenous autonomy, critiques of the state, and questions of migration. It will only continue to become more so as the century unfolds, and the predictable and unpredictable effects of planetary climate change take effect. Thus, it is imperative that as artists and scholars we rise to meet the call put forth by the activist group Futuros Indígenas, who write that "en tiempos de crisis climática, el futuro es un territorio a defender" (in times of climate crisis, the future is a territory to defend).[21]

Each chapter narrows in on a distinct iteration of foreclosure occasioned by extractivist capitalism: toxicity that persists in the wake of transnational mining, violence against land defenders, the sixth great extinction, drought, and the erosion of cultural infrastructure. I look at how authors, filmmakers, artists, and cultural practitioners engage with these legacies of harm and what they mean for the future. By organizing the book around the material foreclosures of extractivist capitalism, I echo the argument made by Verónica Gerber Bicecci, the subject of Chapter 1, who suggests that extractivism is a "dystopian machine."[22] At the same time, I share the hope of Maricela Guerrero, whose poetry I consider in Chapter 3, that the horrors left in capital's wake might yet be reworked to give way to "la posibilidad azarosa de que floreciera algo" (the contingent possibility that something might yet flourish).[23] Through subjunctive tactics like rewriting, counterfactual mourning, contiguity, sensorial immersion, and experiments in green energy, the studied works contest foreclosure and open the future up to renegotiation. This dynamic between foreclosure and subjunctive rejoinder is the fundamental dialectic that organizes my book, which I rehearse in different contexts and texts in order to bring into view the trends that are characteristic of contemporary ecocultural production in Mexico.

FROM DREAMS OF OIL TO POSTEXTRACTIVISM

Subjunctive Aesthetics assesses contemporary Mexican cultural production that responds to the symptoms of climate change and the consequences of extractivist acceleration. In the title, I collapse these two drivers of environmental crisis under the umbrella term "climate change" because, to my mind, the two are inextricable. When we talk about anthropogenic climate change, we are also referring, however implicitly, to the asymmetrical colonial model of extractivism that continues to be the basis of fossil-fueled modernity. We might gloss this continuity, following Horacio Machado, as a civilizational model based on appropriation: the extraction of wealth, nature, and labor from the Global South to facilitate the industrialization and prosperity of the Global North.[24] Climate change is one of many material manifestations of these historical and ongoing asymmetries, evident in the fact that half of global carbon emissions from 1990 to 2015 are attributable to the world's richest 10 percent, while the poorest 50 percent are responsible for only 7 percent of global emissions.[25] Moreover, climate change is the direct result of extractive industries. One study shows that just ninety companies and government-run industries have contributed nearly two-thirds of all global industrial emissions, with Pemex coming in ninth place in terms of cumulative impact.[26] Other environmental effects of extractivist capitalism, like massive deforestation to facilitate monocropping, have further inhibited the planet's ability to recapture carbon. While I am careful to parse the various causes behind specific instances of environmental crisis in each chapter, I take the risk of subsuming these causes under the term "climate change" in the book's title because studying large-scale environmental transformation from the vantage of Mexico insistently links climate change to the systems of extractivist exploitation imposed by colonialism.[27]

Even though the uneven benefits and costs of carbon pollution are symptomatic of capitalist regimes of racialized and spatialized inequity, environmental issues have historically been framed as separate from other sociopolitical problems in Mexico. Such conceptual isolation is in the interest of the state, which has relied upon extraction and the sale of concessions to boost economic growth in the pursuit of what Maristella Svampa has identified as the "developmentalist illusion," or the ever-elusive hope of leveraging natural resources to gain parity with other industrialized countries.[28] This pursuit has made extraction a pillar of state power. Ever since President Lázaro Cárdenas led the radical push to nationalize the energy sector in 1938 by expropriating land, expelling foreign drillers, and founding Pemex, oil has been the material basis of the dream of "resource nationalism," a path to prosperity premised on collective ownership of boundless nature, which is nonetheless risky in terms of its predictable and unpredictable socioenvironmental effects.[29] Historian Germán Vergara argues that fossil fuels are central to understanding the evolution of Mexico over the last century, not only because they "underpinned the longest economic boom in Mexican history . . . averaging 7 percent growth annually between 1940 and 1970," but also because they quickly locked Mexico into a self-fulfilling cycle of "perennial scarcity amidst energy abundance," in which every new influx of oil prompted new demand, with "profound environmental and social consequences."[30] And yet in spite of its material and social failures, the promissory discourse of nature's potential to be converted into capital and, eventually, modernity has continued to legitimize extractivist development policies, as scholars like Charlotte Rogers and Ericka Beckman have shown.[31]

The world's tenth largest oil producer, Pemex pays taxes that since the 1980s have constituted between 30 and 40 percent of the federal spending budget.[32] This means that government programs—including public arts funding—depend heavily on the continued profitability of hydrocarbons, as Marcela Armas's visualization of Mexico as a machine lubricated by oil powerfully illustrates (fig. 0.1). The reliance of national revenue on petroleum, Claudio Lomnitz remarks, is so decisive that Mexican politicians are effectively more indebted to oil interests than to their citizens.[33] It comes as no surprise then that even when the Mexican state has acknowledged environmental crisis, as in a recent SEMARNAT report that acknowledged that Pemex polluted 655 sites in Mexico between 2008–2021, it has perpetually deferred its remediation to an ever-receding future—one that may never materialize, since it would endanger the bottom line.[34] The state's faith in the developmentalist illusion is such that issues like violence, migration, and poverty are often presented as solvable only through more (rather than less) extraction, effectively creating an ouroboros in which environmental exploitation both compounds social vulnerability and is proffered as its solution.

Mexico's petronationalism is supplemented by a classical export-oriented approach to extractivism in other sectors. Booming global mineral prices in the twenty-first century incentivized the right-wing administrations of Vicente Fox and Felipe Calderón, both of the Partído Acción Nacional (PAN, National Action Party), to escalate foreign direct investment in mining. They granted a deluge of concessions "equal to 31 percent of the country's territory" to mining transnationals for rock-bottom prices and with generous terms (no royalties and concession periods of fifty years), making Mexico the site of the greatest foreign direct investment in mining in Latin America.[35] Because these concessions were often granted without community consent, environmental conflicts multiplied. In the face of fierce local opposition to the expropriation of communal lands, the state militarized extractive regions. This escalated regional violence, rendering Mexico one of the most dangerous countries in the world for land defenders, as I discuss in Chapter 2.

The collective payoff of the extractive promise has been ever deferred. There is no "enough" when the name of the game is endless accumulation.

FIGURE 0.1. Marcela Armas, *I-machinarius*, Laboratorio de Arte Alameda, 2008. Armas depicts Mexico as a machine lubricated and powered by oil. By inverting the typical orientation of the map, Armas signals that the gravitational pull of labor and energy produced by the nation flows north, toward the United States. Courtesy of the artist.

As Marxist theorist Bolívar Echeverría has pointed out, capitalist modernity "only actualizes [those possibilities] that promise to be functional for its own goal, the accumulation of capital."[36] Everything else is subsumed to profit, leading to the destruction of place, exacerbation of inequities, exclusion from gains, and erosion of trust. Prominent environmentalist and biologist Víctor Toledo bluntly deems the extractive developmental paradigm a "proyecto de muerte" (death project): a civilizational model that is "la causa profunda, oculta, y principal de la desigualdad social [. . .] así como la mayor amenaza a la supervivencia biológica, ecológica, cultural, y en fin, humana" (the profound, hidden, and principal cause of social inequality . . . as well as the largest threat to biological, ecological, cultural, and human survival).[37] For his part, Echeverría illustrates the "self-defeating" logic of extractivist modernity with a sticky analogy: "It plays an absurd game which, were it not for the blood and tears that it involves, is like a scene from a Charlie Chaplin movie, in which he attempts to climb upward on a downward escalator that is moving faster than himself."[38]

Even today, as oil reserves are dwindling—the Comisión Nacional de Hidrocarburos (National Hydrocarbons Commission) estimates that existing reserves could be depleted as soon as 2030—the conversation among state leaders has shifted to how hydraulic fracturing could allow Mexico to tap into shale reserves estimated to be the world's sixth largest remaining deposit and usher in yet another era of petrofueled development.[39] But the belief that extractivism will actualize a better future is highly contested. It has been problematized by its spotty economic track record, its legacy of social, cultural, and environmental harm, and its causal links to ever-accelerating climate change. (Seventy percent of Mexico's total greenhouse gas emissions since 1960 are attributable to Pemex.[40]) Extractivist projects throughout Mexico are sites of labor disputes, water and air contamination, deforestation, subsidence, and other problems.[41] The question then is not only what *can be* done with territory but the subjunctive problematic of what *should be* done.

Bolívar Echeverría is quick to point out that this is not the only possible version of modernity that could exist. He writes that there is a "potential, virtual" modernity counterposed to capitalist modernity that can still be glimpsed through the cracks of "the actual, empirical, or real."[42] Such a modernity would still aim for abundance but would measure it, not in terms of profit or surplus, but through the flourishing of life. This search for an alternative modernity not circumscribed by ecocidal capital is the basis of a new wave of cultural production that I assess in this book that gestures toward a postextractivist future. This body of work grapples with the illusory promises of extractivism and its many failures but also goes beyond evidence-based critique. It takes up the subjunctive task of imagining how the world might be valued otherwise. Illustrative of this subjunctive dreaming is the Zapatista call for "un mundo donde quepan todos los mundos, tantos mundos como sea necesario para que cada hombre y mujer tenga una vida digna donde sea, y que cada quien esté satisfecho con lo que su concepto de dignidad significa" (a world where all worlds might fit, as many worlds as are necessary so that each man and woman might lead a dignified life wherever they might be, and so that everyone might be satisfied by the meaning of their own concept of dignity).[43]

The quest for a capacious world that can contain many worlds has been robustly put into practice by communities and social movements throughout Mexico, from the successful campaign in 2013 by the Demanda Colectiva en Defensa del Maíz Nativo (Collective Lawsuit in Defense of Native Corn) to halt Bayer-Monsanto's cultivation of genetically modified corn to the ongoing opposition to the Tren Maya megaproject in the Yucatán Peninsula for its potentially ecocidal effects on Indigenous communities and on the Calakmul Biosphere Reserve, Latin America's second largest remaining forest after the Amazon.[44] This book studies how cultural production complements these efforts. How does the ethical imperative to contest extractivism's promissory form play out in aesthetic terms? What narrative techniques are ascendant in the era of renewed ecological imagining? And how should we reckon with the emergence of environmentalist art from within reigning neoliberal structures of cultural production characterized by uneven circulation, corporate funding sources, and commodification?

While environmental concerns have historically been partitioned from other sociopolitical concerns, pervasive anxiety in the twenty-first century about ongoing and future climate change has reached a tipping point. As Carlos Monsiváis concluded in 2005, even though Mexican environmentalism was a relative latecomer (only taking off in the 1980s), its time has finally arrived.[45] This is evident in the outpouring of ecocultural production over the past decade, a body of work that has yet to be comprehensively studied. This boom of environmentally oriented art is not unprecedented, but its quantitative jump merits sustained attention. To mention but one illustrative example, whereas prior to 2010 an average of fewer than ten Mexican films a year touched on environmental topics, that annual rate now hovers around one hundred, a boom I assess in Chapter 4.[46] The exponential explosion of cultural responses to climate change is not limited to any one genre or art form, but ranges from young adult fiction (Alberto Chimal's *La noche en la zona M*) and short stories (Gabriela Damián Miravete's *La canción detrás de todas las cosas*, Andrea Chapela's "Como quien oye llover") to dance (Amanda Piña's *Danzas climáticas*), visual art (Minerva Cuevas's *A Draught of the Blue*), podcasts (*2050: El fin que no fue*), and theater (Giuliana Kiersz, Martha Mega, and Carlos Tavera's *Antes de que suban los mares*).

The quantitative increase in ecocultural production has been accompanied by a qualitative one, a shift in tone that acknowledges the impossibility of thematizing the nonhuman without also foregrounding its state of crisis. The 2019 rerelease of Pablo Soler Frost's book on insects in the Mexican imaginary, *Oriente de los insectos mexicanos*, first published in the mid-1990s, includes an updated epilogue addressing this shift. Soler Frost writes that from the vantage of the present, his original text feels like "una reliquia de otros tiempos, distintos, un objeto que viene de otro siglo" (a relic from another, different time, an object from another century).[47] He explains that his celebration of insects must now be read as elegiac, since many invertebrates are experiencing precipitous population decline or are threatened with extinction. Thus, while pioneering Mexican environmentalist authors and artists, along with countless grassroots activists and Indigenous groups, signaled the dangerously irreversible human and nonhuman costs of development in the mid- to late twentieth century, these harms are now an indisputable reality, bringing with them a necessary reevaluation of the world as we thought we knew it. Soler Frost's retrospective assessment illustrates this ontological scrambling: "El mundo en el que vivimos ya no es el mundo en el que vivíamos. ¿Y el mundo en el que viviremos? Se anuncia ya y sus heraldos son horribles de ver y oír." (The world in which we live is no longer the world in which we used to live. And the world in which we will live? It is announcing itself, and its heralds are terrible to see and hear.)[48]

The crisis of futurity heralded by climate change is acutely felt in Mexico. Studies indicate that Mexicans are more worried about climate change than any other global issue, that they are more anxious about natural disasters than any other quotidian threat, and that suicide rates have steadily risen along with temperatures.[49] Since the 1960s, Mexico has been getting warmer; in 2011 it endured the worst drought on record, which led to the failure of 2.2 million acres of crops, as discussed in Chapter 4. Less hospitable conditions for agriculture have led to food and water insecurity and intensified rural-to-urban migration. Mexico is also one of the deadliest countries in the world for environmental activists: in 2022, thirty-one land defenders were murdered (a third in logging-related conflicts and over half Indigenous), and many more were criminalized or intimidated.[50] Many Mexicans are therefore already experiencing the effects of rising temperatures and ecocidal extraction; climate change is not just a future threat, but one that delimits the present.

ENVIRONMENTALISM IN TWENTY-FIRST CENTURY MEXICO

Building on Monsiváis and Soler Frost's observations that the twenty-first century has seen an undeniable shift in the zeitgeist such that environmental crisis is now a recognizably mainstream point of concern, this study primarily attends to contemporary visual and literary arts. This is not to suggest that ecocultural production did not already exist; to the contrary, in previous work I have argued that anxiety about anthropogenic climate change is perceptible as far back as the speculative musings of Amado Nervo at the turn of the twentieth century.[51] Critics like Gisella Carmona, Jorge Quintana-Navarrete, Kerstin Oloff, Victoria Saramago, Mark Anderson, Laura Barbas-Rhoden, Brian Gollnick, and Micah McKay have similarly excavated ecocritical insight in the works of twentieth-century authors like Alfonso Reyes, José Vasconcelos, Juan Rulfo, Agustín Yáñez, Efraín Bartolomé, Homero Aridjis, and José Emilio Pacheco.[52] Moreover, the 1970s, '80s, and '90s were key decades for the emergence of environmental consciousness in Mexico, centered around polemics such as the construction of the Laguna Verde Nuclear Power Station, the devastating 1985 earthquake, pollution, pesticide regulation, protections for endangered species, and the fight for Indigenous territorial autonomy. Mexico City was ground zero of what Homero Aridjis and Fernando Césarman deemed "urban ecocide," whose lack of regulation Carlos Fuentes posited in *Cristóbal Nonato* (Christopher unborn, 1987) was linked to the state's commitment to development over the health of its citizenry.[53] This period also saw the formation of political advocacy organizations like the Grupo de los Cien (Group of One Hundred), a coalition of artists and intellectuals who aimed to push environmentalism onto the national agenda; the emergence of movements like the Zapatistas, which tied Indigenous political agency to territorial autonomy; and the founding of the conservative so-called "green" political party, Partido Verde Ecologista de México (PVEM, Ecological Green Party of Mexico), which blended conservationism with reactionary and homophobic policies.[54] While my goal is not to track the cultural genealogy of environmentalism in Mexico—a vital task that I hope future scholars will take up—it is important to situate the contemporary emergence of ecocultural production within a longer tradition.

I posit the consolidation of environmentalism in Mexico since 2012 by looking at three factors: institutional, cultural, and legislative. First, the exponential proliferation of ecocultural production mentioned earlier has been accompanied by its institutionalization via the creation of spaces uniquely focused on environmental issues, sponsored by a combination of state, corporate, and nonprofit funding. Exemplary of this expansion in cultural infrastructure is the debut of two specialized environmentalist film festivals: Cinema Planeta, founded in 2009 as a spin-off of Spain's International Environmental Film Festival, and EcoFilm Festival Internacional de Cortometrajes Ambientales (International Environmental Short Film Festival), founded in 2011 to promote environmental shorts. This institutional support for the funding and circulation of environmental films helps explains the explosion of Mexican ecocinema in the last decade studied in Chapters 4 and 5. Similarly, the topic is receiving greater institutional support within Mexican academic publishing, as indicated by El Colegio de México's book series Estudios sobre energía (Energy Studies), inaugurated in 2017. The creation of institutional infrastructure dedicated to environmentalism has heightened its visibility across cultural sectors and brought disparate groups together around a shared question of concern. This can be attributed to the hard work of grassroots organizations like the Asamblea Nacional de Afectados Ambientales (ANAA, National Assembly of Affected Environments), which have pushed broader sociopolitical movements to include environmental issues within their scope. Founded in 2007 by political economist Andrés Barreda Marín, ANAA is a national network that connects communities dealing with environmental damage. This network building has helped erode the conceptual isolation of environmentalism, which, as Gabriela Méndez Cota argues, has long been compounded by the fact that it affects far-flung populations in radically distinct ways, making political coalitions hard to establish.[55]

At the same time that environmentalism has gone mainstream, however, it has become increasingly detached from situated activist efforts. Indeed, as environmental crisis has become more visible across a variety of cultural institutions, it has been presented as a purportedly apolitical issue of collective concern, rather than as a political problem connected to concrete questions of land defense. Thus we have phenomena such as the *México Megadiverso* exhibit installed in 2018 in Mexico City's Chapultepec Zoo, featuring eight enormous *alebrijes* (fantastical sculptures) of endangered animals and promoted by the local news as a great opportunity to take a fun selfie while "reflecting on the natural habitat."[56] Similarly, in 2011 the Museo Interactivo de Economía (MIDE, Interactive Museum of Economics), a private museum founded by the Banco de México, debuted an entire floor dedicated to sustainable development, focused solely on how it pertains to economic growth. FEMSA, the biggest bottler of Coca-Cola products in the world, is the funding source behind the FEMSA Biennial, which has foregrounded environmental themes and has been an important venue for the production of antiextractivist art, like that by Verónica Gerber Bicecci and Naomi Rincón Gallardo analyzed in this book. Such initiatives preemptively highlight these institutions' "green" ethics without troubling their foundations. Echoing the contradictory origins of environmentalism in Mexico as linked to projects as ideologically incompatible as those of the Partido Verde and the Zapatistas, the cultural explosion of apolitical takes indexes the consolidation of societal anxieties around environmental damage, such that a wide array of institutions feel compelled to address it. Consequently, it is worth parsing to what extent the representation of environmental issues entails a true political shift, and to what extent it is merely performative greenwashing, a debate that pops up in various moments throughout the book.

The rise in cultural production dedicated to climate change over the last decade or so also echoes its legal inscription by the state. In 2007, in the midst of catastrophic flooding in the state of Tabasco, Mexico officially recognized climate change as a major national challenge, launching its National Climate Change Strategy. The following year, it developed short-term emissions reduction targets through the Special Program on Climate Change and passed the first legislation explicitly targeting climate change through renewable energy use. In 2010, Mexico positioned itself as an international leader on this issue by hosting the UN Climate Change Conference (COP16) in Cancún; in 2012, it adopted the UN's General Law on Climate Change with the goal of halving national emissions by 2050. In 2015, Mexico put in place an Energy Transition Law, aspiring to generate 35 percent of the country's energy through clean sources by 2024. And so on. Although many of these moves were merely symbolic, they nonetheless bracket a new chapter in which dealing with climate change has been legally codified as crucial to Mexico's present and future.

As in other countries around the world, the legal recognition of the need to confront climate change has existed at odds with official state policies, which continue to promote extractivist development. The establishment of targets rather than mandates means that implementation and enforcement have remained conveniently slippery. Such contradictory posturing transcends political affiliation. PAN President Felipe Calderón positioned himself as Mexico's first environmentalist president, peppering his policies with green discourse, while at the same time ushering in open-pit mining and coastal development projects that devastated 75 percent of the country's mangrove forests.[57] His successor, PRI President Enrique Peña Nieto, fashioned Mexico as an international climate leader in the UN Paris Agreement while privatizing Mexico's energy sector. Current left-leaning MORENA President Andrés Manuel López Obrador has also paid lip service to ecological issues at the same time that he has dismissed Indigenous land defenders as "far-left radicals."[58] Put simply, climate change has been acknowledged across the political spectrum in Mexico as a key issue, and yet this acknowledgment exists in uneasy tension with state reliance on petroleum and the continued ascription to the colonial-capitalist maxim of economic growth as the marker of state success.

None of this, of course, is happening within a geopolitical vacuum. Trade agreements between the US and Mexico, like NAFTA and USMCA, have encouraged US companies to site factories in Mexico, where labor is cheaper and regulations less enforceable. Corporate environmental damage is promoted by NAFTA itself, which does not include bilateral measures to hold transnational companies accountable for their environmental footprint. This creates a situation in which Mexico, dependent on these corporations for jobs and economic stability, has little incentive or capability to punish transnational offenders outside of imposing insignificant fines. Meanwhile, the toll of these polluters is borne by Mexican territories and peoples, as dramatized in the 2006 documentary *Maquilápolis* (City of factories, directed by Vicky Funari and Sergio de la Torre). *Maquilápolis* illustrates Fernando Coronil's forceful argument that the commodification of territory and labor go together: their cheap availability is the foundation of a nation's potential wealth. The present-day dynamic instantiated by NAFTA merely furthers the colonial "international division of nature," in which the cost of developing the Global North is displaced south.[59]

This book joins the efforts of scholars across fields who have argued that Global South perspectives are crucial to understanding the planetary dynamics of ecological collapse. Against the disingenuous abstraction of the term *Anthropocene*, the current geological age defined by human activities, whose prefix, *anthropo-*, invokes a disembodied, singular human experience that subsumes the entire species within the logic of Western colonial capitalism, Latin American accounts of the Anthropocene argue that the practices that have driven climate change are grounded in time and space.[60]

Inscribed by explorers like Alexander von Humboldt as an empty expanse of untouched wilderness, Latin America unfurled in the European imagination as a ready-made site for extraction.[61] This laid the ideological groundwork for today's globalized neo-extractive economies, reliant upon unequal exchange and limitless, cheap nature and labor. If Latin America has been historically framed as belatedly experiencing modernity, a lag that Latin American nation-states hoped to remedy by leveraging their natural resources, this developmentalist approach has indeed catalyzed a temporal reorganization, but not in the way expected. Rather, as Amitav Ghosh glosses, "the Anthropocene has reversed the temporal order of modernity: those at the margins are now the first to experience the future that awaits us all."[62] The documentaries studied in Chapter 4 likewise index that rural Mexico is already experiencing the heightened conditions of drought and flooding that presage the precarious future in store for the entire planet.

Latin America not only portends the future of the Anthropocene, it is also its site of origination. While many scientists have tied the Anthropocene to the Industrial Revolution, Simon Lewis and Mark Maslin have persuasively argued that the Anthropocene began in 1610 with the European colonization of the Americas.[63] The genocidal eradication of over 55 million native peoples (via enslavement, murder, disease, overwork, and famine) ground traditional farming to a halt and consequently prompted the reforestation of over 55 million hectares of forest. This massive reforestation caused a sharp drop in atmospheric carbon dioxide in 1610, known as the Orbis spike, that is observable in Antarctic ice cores. The decimation of Indigenous peoples had global climatic reverberations, contributing to an era of global cooling known as the Little Ice Age.[64] Genocide operationalized extraction: emptying the landscape so that it could be appropriated and rewritten by colonizers.[65] Agricultural lands became pastoral ones, and newly introduced European livestock and plants catalyzed desertification.⁶⁶ In colonial Mexico, this ontological-epistemological unworlding was leveled against the unique hydraulic agricultural system of urban Tenochtitlan, the Aztec capital, whose water system was destroyed so that the Valley of Mexico could be rebuilt in Spain's image.⁶⁷ Turning our attention to Latin America reminds us that the analysis of present-day ecological collapse necessitates a critical framework expansive enough to account for global dynamics of empire and race.

A NEW STRUCTURE OF FEELING

Over the last decade, anxieties about ongoing and future climate change have percolated into what Raymond Williams calls a "structure of feeling," or a social experience of the present that is still in process, provisional, and unresolved. Williams defines structures of feeling as forming in opposition to hegemonic iterations of power but not yet fully fixed in their orientation. According to Williams, art and literature are often where we can find "the very first indications that a new structure is forming."⁶⁸ Subjunctive aesthetics reflect the paradox that in spite of knowing with greater certainty that extractivist capitalism and fossil-fueled modernity are and will continue to destructively reshape the planet, this knowledge has not yet translated to collective action, but to greater uncertainty vis-a-vis the future, generating a zeitgeist characterized by anxiety, doubt, and disillusion.

In times of collective anxiety, the subjunctive mood flourishes because it is the grammar of uncertainty. Elaborating this point, Barbie Zelizer writes that the subjunctive's "concomitant invocation of emotionality, contingency, and imagination ... becomes particularly useful around events that are unsettled, ambiguous, difficult, contested, or in otherwise need of public consensus," by asking us to feel or imagine our way through them.⁶⁹ I can think of no better description for the climate crisis. Anthropogenic climate change has troubled what Anthony Giddens calls "ontological security," or the sense of safety derived from the stable continuity of our social and material environments.⁷⁰ It foregrounds the contingency that has always been a part of the planetary ecosystem but that has been subsumed into predictability: the neat division of rainy and dry seasons or the progressive glide from winter into spring.

As a grammatical mood, the subjunctive gives form to a state of responsive suspension and models relationality, the way that subjects exist in contingent relation, in response, reliance, or subordination to one another. Extending these observations to the theorization of the subjunctive as an aesthetic mode, subjunctive aesthetics are produced from a relational position conditioned by historical determinants and power structures that frame *what is*—atmospheric conditions borne of settler-colonial accumulation, economic models in which art, biennials, and museums are funded by extractivism, and so on. Yet from within this position of contingent dependence, subjunctive rejoinders to the status quo nonetheless exert a form of cultural agency that we might deem "modest" following Claudio Lomnitz, in that they register and reflect the paradoxes and ambiguities inherent to their production. Without becoming paralyzed by entrenched inequities of power that seem to proffer no exit, subjunctive aesthetics illustrate the capacity of imaginative work to intervene in everyday thought and action, "to instantiate, to help press, express, and at times redress contradiction."[71] This imaginative work of cultural production does not produce the same outcomes as the work of activism and organizing, imperative to actualizing concrete political change. But what art can do is reshape shared narratives and desires through conceptual forms that model relational territorial modes.

In *Weak Planet*, Wai Chee Dimock similarly describes the subjunctive as a "crisis-necessitated form."[72] A state of responsive suspension that strains against what is, the subjunctive mood performs a temporal trick: it "enables those who don't have much time left to spend it without reservation and without panic, as if an infinite future were still at their disposal."[73] Subjunctive imagining, in Dimock's assessment, is not concerned with outcomes or even whether or not something is likely to happen, but with the possible. In the face of potentially irreversible environmental foreclosure, the improvisational and inconclusive character of subjunctive imagining is a necessary complement to the urgency of political action, because it is uninhibited by pragmatism. The Chilean Anthropocene manifesto published in 2019 illustrates this point in its claim that the Anthropocene should not be seen as "a closed scientific category to be accepted" but as "a space of debate ... a call to invent new possible futures."[74] This manifesto is just the latest in a long history of Latin American subjunctive appeals to redefine the relationship with territory. Mexico has a particularly rich history of such imaginative reconfigurations, ranging from Emiliano Zapata's 1911 Plan de Ayala, with its behest for equitable land redistribution, to the more recent Zapatista clamor for "un mundo donde quepan muchos mundos," an appeal to reimagine the planet as ontologically plural.

By tying the theoretical underpinning of this book to a grammatical mode, I make a conscious argument about the power and possibility of language in responding to climate change. Even though the effects of climate change take place outside of language, it is through language that we either reify the status quo or seek to reimagine it. Language is how we communicate our experience of the world; it also shapes our perception of it. Subjunctive aesthetics open up a space from which reality can be questioned and from which we can speculate about other possibilities; it makes other realities seem plausible by giving us a means to first imagine and then utter those possibilities aloud.

CHAPTER SUMMARIES

The book's first three chapters identify different subjunctive aesthetic strategies that are used to contest ecocidal foreclosure: rewriting, counterfactual imagination, and contiguity. Chapter 1 traces the boom of environmental rewriting as a genre of Mexican literature that mirrors the revisionary logic of climate change. After a prelude centered on Alfonso Reyes's recantation of his prior celebration of the desiccation of the Valley of Mexico, the chapter focuses on works of environmental rewriting by visual artist and author Verónica Gerber Bicecci. In *Otro día . . . (poemas sintéticos)* (Another day . . . [Synthetic poems]) and *La compañía* (The company), Gerber develops a method of environmental rewriting through the inhabitation of extant textual forms. By dwelling within texts inherited from José Juan Tablada and Amparo Dávila and reworking them from within, Gerber demonstrates the renovative potential of composting past forms and propelling them into the future.

Chapter 2 turns to the high rates of violence experienced by Indigenous land defenders. It suggests that ecocriticism's outsized emphasis on slow violence has obscured the formative role of bare violence in the implementation of the extractivist economy. The chapter explores how land defenders are mourned, with particular attention to counterfactual strategies of mourning such as we see in activist slogans like "Samir vive" (Samir lives) that negate the finality of death. I contend that these counterfactual strategies of mourning are less interested in bringing these cases to legal resolution and more invested in inviting the public to conceptualize territory outside of profit. The chapter then turns to Naomi Rincón Gallardo's performance trilogy *Una trilogía de cuevas* (A cave trilogy), in which the speculative enactment of murdered Mixtec activist Bety Cariño's descent through the underworld fleshes out the utopian counterworlds that land defenders continue to nourish even in death.

Chapter 3 attends to another manifestation of colonial-capitalist foreclosure: species extinction. It studies how four Mexican poets have thematized extinction and animal endangerment, departing from a reflexive critique of how representation has transformed the extinct into what Isabel Zapata terms a "phantom limb." Karen Villeda in *Dodo* reorients the fascination with the extinct back to the system of imperial expansion that eradicated it. This materialist approach to the world is furthered by Xitlálitl Rodríguez Mendoza's *Jaws*, which elucidates the analogous commodification of human and nonhuman life through contiguity. Poetry's formal ability to bring different things and beings into proximity within the delimited confines of a compressed space is shared by Maricela Guerrero's *El sueño de toda célula* (The dream of every cell), which inscribes the subjunctive yearning for a language outside of empire's taxonomical drive.

In the fourth and fifth chapters I turn to cinema and rural Mexico in the era of climate change. Here my attention shifts from the subjunctive's utopian potential to its operation as a subjoined structure, to think through how existing structures of cinematic funding, distribution, and exhibition shape the meaning produced by ecocinema, or films with environmentalist content. Chapter 4 studies the surge in contemporary documentaries that thematize rural resilience. It looks at three sensorial ethnographies: Everardo González's *Cuates de Australia* (Drought), Betzabé García's *Los reyes del pueblo que no existe* (Kings of nowhere), and Laura Herrero Garvín's *El Remolino* (The swirl), which all focus on areas that have been rendered seemingly unlivable by the scarcity or overabundance of water. These films offer fertile terrain for thinking about the contradictory meanings of resilience in the twenty-first century, both as a frame for celebrating the persistence of life amid conditions that others would deem inoperable and as a problematic neoliberal displacement of the responsibility to adapt to crisis away from the state and back onto the individual. These tensions come into heightened relief, I suggest, as these films about rural resilience circulate and are consumed by urban transnational publics.

Chapter 5 further considers the uneven circulation of Mexican cinema, highlighting the erosion of exhibition infrastructures in rural Mexico. The discussion of infrastructure presents me with an opportunity for a methodological pivot to further consider the conditions of production and consumption that define contemporary culture, as well as their imbrication with extractivism. A case study of Cine Móvil ToTo, an experiment in ambulatory cinema powered by bicycles and solar panels, leads me to think in the subjunctive about our desires for the future of cinema.

Collectively, the works considered in *Subjunctive Aesthetics* signal a renewed investment in the intersection of art and territory: art as a space in which to think imaginatively about the value of place beyond profit and as a space from which to respond to the seeming foreclosure posed by extractivist capitalism that Mark Fisher has described as the "slow cancellation of the future."[75] What unites these projects, I propose, is their activation of the subjunctive "as if" to trouble normative assumptions about extractivism, reveal its illusory logic and outrageous inequities, and signal other ways of being in relation to territory. This book contends that by considering the subjunctive not just as a grammatical mood but as an aesthetic modality, we might better understand how cultural responses to climate change help dispute the current reality as the only option and imagine alternative paths that might engender better futures. By thinking about these questions through the subjunctive, I foreground the role of desire and imagination in producing the future. Dominic Boyer has observed that desire typically models itself on the pleasures of the past and "will do anything in its power to pull the future into the gravitational orbit of the past."[76] Rather than chase the impossible dream of replicating past fossil-fueled growth, how can art cultivate alternative desires that might ensure the future of the territorial commons? This is the work of subjunctive aesthetics.

CHAPTER 1
Environmental Rewriting

The Anthropocene is a revisionary mode: the anthropogenic amending and rewiring of the planet's basic functioning. Alfonso Reyes arrived at this conclusion *avant la lettre* in his foundational essay, *Visión de Anáhuac (1519)*. Penned from exile in Madrid in 1915, during the Mexican Revolution, Reyes proposed that what united Mexico's precolonial and colonial past with the postcolonial present was the shared transformation of the Valley of Mexico, known in Nahuatl as Anáhuac, from an aqueous landscape into the land-bound metropolis of Mexico City. For Reyes, Lake Texcoco's desiccation conjoined Mexican identity: process and proof of the joint trajectory of the peoples that inhabited it.[1]

Anáhuac's landscape, molded by human ambition, was for Reyes the site of Mexico's "national essence" and its pristine air, its defining feature. He snappily deemed it "la región más transparente del aire" (the air's most transparent region), in a riff on Alexander von Humboldt's observation a century earlier about the bodily pleasure that the valley's topography inspired.[2] The confluence of altitude, lakes, and surrounding mountains endowed Anáhuac with a consistently cool climate, celebrated by Humboldt as "otoño perrene" (perennial autumn).[3] The catchy phrase "la región más transparente del aire" effectively marketed the region's unique sensibility and soon became omnipresent in advertising, journalism, and literature about Mexico City, culminating in Carlos Fuentes's eponymous novel *La región más transparente* (1958).

In 1940, thirty-five years after he published *Visión*, Reyes lamented the unexpected upshot of the valley's anthropogenic transformation: dust clouds that obscured the sun and viciously whipped through the city.[4] He angrily mused in the *crónica* "Palinodia del polvo" (Dust Recantation) that the region's desiccation and urbanization—the very practices he had celebrated as grounding state progression in *Visión*—had ruined its air quality. Reyes opened the recantation by invoking his famous phrase, now as an accusatory interrogative:

> ¿Es esta la región más transparente del aire? ¿Qué habéis hecho, entonces, de mi alto valle metafísico? ¿Por qué se empaña, por qué se amarillece? Corren sobre él como fuegos fatuos los remolinillos de tierra. Caen sobre él los mantos de sepia, que roban profundidad al paisaje y precipitan en un solo plano espectral lejanías y cercanías, dando a sus rasgos y colores la irrealidad de una calcomanía grotesca, de una estampa vieja, artificial, de una hoja prematuramente marchita.
>
> Is this really the region of the clearest air? What have you done, then, with my high metaphysical valley? Why is it tarnished, why does it turn yellow? Whirlwinds of earth rush over it like ignis fatuus. Like a cloak, sepia tones steal all depth from the landscape and precipitate a single spectral plane far and near, giving its features and colors the unreality of a grotesque decal, an old, artificial print, a prematurely withered leaf.[5]

For Reyes the dust storms posed an aesthetic problem—the landscape's loss of depth and color—even more than a problem for human health. It was also a problem of authorship. Minute particles, not human design dreams, were the valley's true organizing force: "El polvo es el alfa y el omega. ¿Y si fuera el verdadero dios? Acaso el polvo sea el tiempo mismo, sustentáculo de la conciencia." (Dust is alpha and omega. And if it were the true God? Perhaps dust might be time itself, fulcrum of consciousness.)[6] Metonymic of this loss of authorial control over the valley's aesthetics was the loss of control over the meaning of Reyes's iconic phrase. In the context of annual dust storms, the saying "la región del aire más transparente" had become a mockery of itself.

As dust storms gave way to dense smog that clung to the industrializing valley in the following decades, others underscored the irony of the phrase. In the 1960s, "la región del aire más transparente" appeared in numerous headlines about Mexico City's air pollution. One op-ed published in *Jueves de Excelsior* in 1968 sarcastically noted that a person would have to be delusional to utter it with a straight face.[7] The fate of Reyes's slogan was to become a sardonic antislogan. This refiguring added yet another revisionary layer to *Visión de Anáhuac (1519)*, which itself was a re-visioning of the past, evidenced by the playful titular inversion of 1915, the year Reyes wrote the essay.

The trajectory from *Visión* to "Palinodia," from design dreams to recantation, illustrates how the revisionary project of anthropogenic environmental transformation, when confronted with its ecological consequences, compels revisionary literary modes.[8] The recantation is one such mode. Derived from the Latin, *recantare* means "to sing in reverse"; the word *palinodia* similarly encodes this idea of return in song through the refrain. The recantation or palinode signifies both repetition and difference, turning back to the past with an eye toward rectifying or recoding the present. The gesture of turning back to write anew figures any given text as undecided and open to future change. Revisionary rewriting is thus a literary method that enacts the temporal simultaneity demanded by the Anthropocene: a looking back to the past while thinking with the present and future. This simultaneity is evidenced in the way a singular phrase like "la región del aire más transparente" folds back in on itself, at once indexing the specter of Lake Texcoco, Humboldt's imperial gaze, Reyes's nationalist dreams, the dust-filled urbanization of the mid-twentieth century, and the contaminated future of a petrofueled city nestled in smog and grappling with water scarcity.

In recent decades, rewriting has become a salient tactic in contemporary Mexican literature about environmental crisis. This revisionary turn takes up the mantle of Reyes's recantation, echoing its blend of outrage, lament, and prophecy. I open with Reyes to orient our understanding of revisionary environmental writing as nothing particularly new. As Lesley Wylie demonstrates, the imperative to rewrite colonial inscriptions of American landscapes was kicked off by the *novela de la selva* of the early to mid-twentieth century. Rather than turn away from European takes on the tropics, for Spanish American authors, rewriting was a way to "contest and supplant disabling cultural stereotypes" and, in turn, reinvent "tropical landscape aesthetics."[9] This revisionary impulse was often discredited as derivative, but Wylie contends that the telluric genre's "imitative proclivities" actually ground a postcolonial method of authorial reclamation to write about the tropics from the tropics.[10]

Building off Wylie's study, this chapter posits that nearly one hundred years after the boom of the novela de la selva, another round of environmental rewriting has taken off. This time, the rewriting is compelled, not by the need to contest tropes and reclaim authorial control over landscape aesthetics, but rather by the ethical imperative to relinquish the fantasy of individual authorial control and embrace planetary rewriting as a collective task. As a subjunctive strategy, rewriting dramatizes the affordances and constraints of inherited form and signals that subordination to extant forms does not prevent their contingent re-vision.

This chapter explores environmental rewriting as a narrative strategy and method through the work of visual artist and author Verónica Gerber Bicecci. In 2019, Gerber Bicecci published two works that originally debuted as museum installations: *Otro día . . . (poemas sintéticos)* (Another Day . . . [Synthetic Poems]) and *La compañía* (The Company). Both intensively rework twentieth-century literary texts by José Juan Tablada and Amparo Dávila, respectively. Unlike other revisionary, intertextual, and archival forms of rewriting popularized since the 1990s, Gerber Bicecci's engagement with her source texts is programmatic. She replicates their formal infrastructure and inhabits them, reworking them from within, altering their content to varying degrees to speak to ecological harm.

Gerber Bicecci uses rewriting to approximate the expansive, sedimented temporalities of the Anthropocene—processes that are long in the making and collectively produced. A single literary object can only emulate the earth's unwieldy temporalities and capacious collectivities, Gerber Bicecci's work suggests, by similarly straddling past and future in method and form. *Otro día . . .* and *La compañía* enact this temporal simultaneity by inhabiting textual predecessors and revising their logic to bear witness to the present and future implications of extractivist capitalism. Rewriting is thus not just backward looking, as it appears in Reyes's recantation, but forward looking, speculative in effect. The loss of authorial control produced by Gerber Bicecci's intensive approach to rewriting moves writing away from an individual practice and toward a process of collective production and sedimentation, mapping onto Cristina Rivera Garza's description of "geological writings," or works composed of countless layers "of connection to language as mediated by others' bodies and experiences."[11] This collaborative ethos models the importance of collective action in the face of shared crises like climate change and extractivist harm.

In what follows, first I touch on a general overview of environmental rewriting in contemporary Mexican literature. Then, I turn to Verónica Gerber Bicecci's *Otro día . . .* and *La compañía* in order to unpack how rewriting operates as a subjunctive method that approximates the disjunctive simultaneity of the Anthropocene. By taking a text out of time and speculatively projecting it into the future, rewriting activates the "as if" that animates the subjunctive as a mode of critical thought and interpretation.

ENVIRONMENTAL REWRITING

Rewriting is an invitation to reconsider with potentially transformative formal, political, and ideological implications.[12] It is an opportunity to think differently about something or to question how things have been done. But rewriting is not a clean break. It acknowledges indebtedness to the past. *Rewriting* is an imprecise term that can mean many things, and indeed it has been used to describe the entirety of literature, as Jorge Luis Borges proposed in "Pierre Menard, autor del *Quijote*." Roland Barthes globally noted in "The Death of the Author" that "a text is made up of multiple writings," and Julia Kristeva, glossing Mikhail Bakhtin, observed that intertextuality lies at the root of all literary meaning.[13]

Narrowing the scope to more overt, intensive forms of rewriting that emerged in the 1990s, Christian Moraru's study of postmodern rewriting tracks the transition of rewriting from an operative mode of "underwriting" ("support and duplication of the already-written") to "counterwriting," which Moraru defines as a form of narrative refiguring that establishes ideological distance from both the "literary past" and "various hegemonic narratives active at the moment of . . . 'redoing.'"[14] Akin to the act of refurbishing a historic building by repairing, repainting, and redecorating it, the intensive rewriter occupies entrenched cultural or political narratives and reworks them from within. The act of retelling activates contradictory outcomes. On the one hand, it further cements the source text's canonicity, extending it into the future. On the other, oppositional retellings estrange well-known narratives and expose their political implications.

Leaning on Moraru, I define environmental rewriting as a practice of artistic production through reproduction that turns back to past texts with an eye toward reorienting them to bring forward their ecological stakes. A loose corpus of Mexican environmental rewriting nascent in the twenty-first century includes recent novels like Yuri Herrera's *El incendio de la mina de El Bordo* (2018; translated as *A Silent Fury: The El Bordo Mine Fire* in 2020), which revisits the recorded history of a mining disaster in order to make reparative sense of the archive's silences; Francisco Antonio León Cuervo's *Nu pama pama nzhogú / El eterno retorno* (Eternal Return, 2019), which remixes Mazahua oral histories of late nineteenth-century Zacatón grass cultivation; and Cristina Rivera Garza's *Autobiografía del algodón* (The Autobiography of Cotton, 2020), which reimagines the story behind José Revueltas's *El luto humano* in tandem with her own familial history in the cotton-growing north. Within this broadly defined corpus of environmental rewriting we might also include books of poetry like those analyzed in Chapter 3, Xitlálitl Rodríguez Mendoza's parodic engagement with the 1970s thriller *Jaws* and Karen Villeda's critical reframing of nineteenth-century novelizations of the dodo bird. We might also consider the genre of environmental rewriting through visual projects like Laureana Toledo's "Not with a Bang" (2015), which sutures photographs taken by Weetman Pearson (the contractor who designed the Trans-Isthmus Tehuantepec Railway for

Porfirio Díaz) with contemporary images of wind farms to underscore the continuity of imposed developmental dreams on the Isthmus of Tehuantepec; or Nicolás Rojas's short film *Tuyuku / Ahuehuete* (2019), which interweaves oral accounts of Mixtec origins around a shared threatened landscape.

Rewriting drives many projects about environmental crisis because of its temporal capaciousness. Rewriting is a method of artistic production that approximates the timescales of environmental crisis, which are not necessarily linear but rather are layered, sedimented, recursive, and disjointed. The temporal work of disjuncture and simultaneity activated by rewriting is akin to the dialectical method advocated for by Walter Benjamin in *The Arcades Project*. Excavating a past work to consider it in discontinuous relation to the present, Benjamin writes, facilitates the "telescoping of the past through the present."[15] Like the telescope, a work of rewriting makes objects distant in time and space appear more proximate to the viewer. This "dialectical penetration," Benjamin explains, "puts the truth of all present action to the test" and "serves to ignite the explosive materials that are latent in what has been."[16] Applied to environmental rewriting, Benjamin's observations underscore that rewriting is not just a corrective endeavor that aims to amend the past but also a disjunctive encounter with the past that animates it in potentially unforeseen ways. At the same time, this encounter with time out of time applies pressure to our understanding of the present. It allows us to consider from the vantage of the present the *longue durée* of planetary history.[17]

More so than in any other example of contemporary environmental rewriting, Gerber Bicecci enacts Benjamin's dialectical method through her strict occupation of inherited literary form, establishing a parallelism between the work of art and the planet as a commons. Like the commons, the work of art as Gerber Bicecci positions it is intelligible as a totality yet internally malleable and modifiable, composed of an assemblage of interacting parts. The meaning of art, then, is not manifest in the individual piece but in the series of interpenetrating pieces accumulated over time.

WHAT TABLADA WOULD HAVE WRITTEN: *OTRO DÍA . . . (POEMAS SINTÉTICOS)*

Unlike other forms of intertextuality that wink at narrative precursors without naming or replicating them, Gerber Bicecci's rewriting is programmatic. Both *Otro día . . .* and *La compañía* explicitly acknowledge their source materials, so the reader is aware that the resulting texts are the product of speculative transtemporal collaboration. In *Otro día . . .* this nod comes in the prologue; in *La compañía* it is mentioned in the appendix. In both cases, she borrows the original texts' narrative logic and proceeds to rework their meaning from within while leaving their formal infrastructure intact. Such a praxis of revisionary habitation metonymically enacts dwelling in the era of climate change, an act that requires juggling the constraints of inherited infrastructure and discourse and the need to relationally reimagine extant systems of thought.

This exercise puts into practice the set theory posited in Gerber Bicecci's best-known work, the novel *Conjunto vacío* (Empty Set, 2015). A branch of mathematical logic that arranges the world into collections of things, set theory draws links between disparate entities, organizing and reorganizing them into an infinite set of visible and invisible combinations. Set theory's networked interpretation of the world has political implications. In *Conjunto vacío*, the narrator explains that during Argentina's Dirty War, set theory was banned from public education. She hypothesizes that this censorship reflects the authoritarian consolidation of power through isolating strategies of "dispersal: separation, scattering, disunity, disappearance."[18] By contrast, to imagine and conceptualize sets is "to form communities, to reflect collectively, to discover the contradictions of language, of the system. Visualized in this way, 'from above,' the world reveals relationships and functions that are not completely obvious."[19] Gerber Bicecci's rewriting of Tablada and Dávila similarly creates sets, forging bonds between past and present works that together make sense of the human relationship to the nonhuman world.

Gerber Bicecci puts set theory's tenets of collective ideation and generative relationality into practice in *Otro día . . . (poemas sintéticos)*. *Otro día . . .* debuted as an installation in the New York art gallery PROXYCO in November 2017 in the *Talon Rouge* exhibit curated by Daniel Garza Usabiaga. The gallery commissioned six Mexican artists to revisit José Juan Tablada's oeuvre and create pieces inspired by his work. The results were exhibited next to a selection of Tablada originals, a dialogic format meant to echo the "diplomatic exhibitions of Mexican art during the twentieth century" that displayed pre-Columbian artworks alongside those of avant-garde modernists to signal continuities in Mexican aesthetics.[20] Gerber Bicecci subsequently published *Otro día . . .* as a book with independent publishing house Almadía in 2019, one hundred years after Tablada's original.

Gerber Bicecci's work straddles the traditionally siloed sites of the museum installation and the book. *La compañía*, which I discuss ahead, similarly debuted as an installation in the Museo de Arte Abstracto Manuel

Felguérez as part of the 2018 FEMSA Biennial, themed "Nunca fuimos contemporáneos" (We were never contemporary), before it was published as a book. Similarly, *En el ojo de Bambi* (In the eye of Bambi, 2020) reimagines Felix Salten's *Bambi: A Life in the Woods* (1928) both in book form and through a curated exhibit of works from Colección "la Caixa" for Whitechapel Gallery.[21] Gerber Bicecci's most recent project, *La tierra es plana como una hoja (y cabalga en el aire)* (The earth is as flat as a sheet of paper [and rides in the air], 2021), takes environmental rewriting to its most literal conclusion, using superimposed layers of paper appropriated from geological surveys overlaid with drawings to create a book object that operates like palimpsestic strata.[22]

A self-described "visual artist who writes," Gerber Bicecci explains that she uses "the book in the same way [as] a wall or a performance," that is, as an exhibition space.[23] Her artistic practice migrates across sites and form, blending text and image without privileging one over the other. The book objects she produces retain intermedial elements that evidence their previous instantiation as exhibits: the dialectic combination of text and image, as well as the use of blank space on the page to emulate a gallery wall. While this intermedial, hybrid combination of script and image sets Gerber Bicecci apart from her peers, it is a strategy characteristic of experimental poetry that dates to the twentieth century.[24] According to Willard Bohn, Tablada was among the first Latin American poets to promote visual poetry. By synthesizing text and image, Tablada moved poetry away from narrative rigidity and toward a more playful, dynamic, and "modern" mode of expression.[25] He would go on to experiment with pictographic poetic forms like the calligram, finding inspiration in the Chinese ideogram and French poetry.

In an interview with Perla Velázquez, Gerber Bicecci describes *Otro día . . .* as a "critical homage" to Tablada's book of haikus *Un día . . . (poemas sintéticos)* (One Day . . . [Synthetic Poems], 1919).[26] She explains that her revision was motivated by wondering how Tablada's book would have been written differently if it were penned today, a century after its initial release. Read within the context of pervasive ecological crisis, Tablada's celebration of the timeless immutability of the nonhuman in *Un día . . .* is scrambled: it is shown to be contingent upon climatic stability. This discordance is playfully indexed by Gerber Bicecci's alteration of the book's title from *un día* to *otro día*. Whereas Tablada's use of the indefinite article situates his observations as taking place on an unspecified day that can stand in for any other, Gerber Bicecci's *otro día* foregrounds sequential subordination (tying this other day to those that have come before it) and contingency (the variability of this *other* day), defining characteristics of the subjunctive grammatical mood. This titular modification echoes the temporal ambiguity of Tablada's original title but also contests its suggestion that environmental timescales are defined by their interchangeability.

Tablada's *Un día . . . (poemas sintéticos)* was published in Caracas in 1919, where he lived in exile in the wake of the Mexican Revolution.[27] While abroad, Tablada continued to advance his intellectual project, which aimed to renovate Mexican aesthetics by combining popular culture with novel aesthetic forms. He was inspired by *japonisme*; in Japanese aesthetics Tablada found the ideal model for modernizing Mexican literature.[28] Japan had seemingly resolved the age-old opposition between "primitive" non-Western tradition and economic growth, understood at the time to be mutually exclusive. It promoted itself as an international "symbol of a unique blend of tradition and modernity," a blend that Tablada and others felt that Mexico could emulate.[29] Western scholars reinforced this affinity, grouping Mexico and Japan as similar sites of cultural exoticism.[30] Tablada felt that if Japan could be celebrated and fetishized worldwide for its unique aesthetic and spiritual traditions, so too could Mexico.[31]

In his search for a modern form, Tablada turned to the Japanese haiku, whose compact structure and direct simplicity departed from the verbosity of *modernista* poetics. Tablada quite literally imported the haiku's tenants: *Un día . . .* was the first book of haikus ever published in Spanish. Structured around four times of day—morning, afternoon, dusk, and night—*Un día . . .* contains thirty-seven poems, each dedicated to a specific animal or plant (and, in one case, a hotel), in the style of a bestiary.[32] Each poem is accompanied by an original watercolor illustration in a circular format reminiscent of Japanese *ensō*, or ring-shaped ink painting. In a departure from tradition, each haiku is titled.[33] Nor did Tablada consistently follow the haiku's rigid seventeen-syllable limit, traditionally arranged in three lines following a 5–7–5 syllable count. More important than these formal specifications for Tablada was the haiku's concise, observational style. According to Bashō, the Japanese poet to whom Tablada dedicated the volume, the haiku materializes an enchanted encounter: immersion within a singular moment.[34] In the book's prologue, Tablada describes the art of the haiku as the practice of capture, comparing it to pinning butterflies or drying herbs.[35] For Tablada, the poem functions like a flower press or time machine: a way of preserving a fleeting moment, so that it might be revisited in the future.

Many of Tablada's haikus were themselves rewritings. Seiko Ota has demonstrated that several of Tablada's poems in *Un día* were based on canonical haikus that the Mexican poet would have first encountered in

A History of Japanese Literature, a book of Japanese poetry translated into English by W. G. Aston in 1899. (In spite of his Japonism, Tablada could not read Japanese.) The poem "Hojas secas" liberally paraphrases an eighteenth-century haiku by Yokoi Yayū. Others more selectively appropriate imagery from Aston's translation. The evocative description of the butterfly as a dry leaf in "Mariposa nocturna," for instance, is borrowed from a haiku by early-sixteenth-century poet Arakida Moritake.[36] *Un día . . .* 's appropriation of Japanese tradition, as mediated by Aston's translation, indicates that *Un día . . .* was already a palimpsestic text.

By rewriting Tablada's rewriting, Gerber Bicecci enters into this circuit of repetitions and reproductions, extending these efforts in time and space. *Otro día . . .* reinscribes the transnational map that connects Japan and Caracas with Mexico via the United States—in this case, New York, the site of Gerber Bicecci's gallery exhibit at PROXYCO. But whereas Tablada's rewriting was motivated by the desire to update nationalist aesthetics, Gerber Bicecci's project is motivated by the search for an artistic method that can approximate the timescales of climate crisis and is informed by the collaborative ethos that this crisis requires.

In a gesture akin to Tablada's gleaning of content from Aston's translations, in the prologue to *Otro día . . .* , Gerber Bicecci explains that she sourced the content of her haikus from Google searches about the subjects. Consequently, her haikus are not uniform in approach. Some allegorically invoke their titular animals, like the cold-blooded toads that push the nuclear button in "Los sapos" or the buzzards that cruelly erect border walls in "Los zopilotes." Others have a pedagogical slant, offering lessons for times of crisis, such as how to propagate a willow by cutting off a healthy branch in "El sauz." Others present opaque omens, like the heron's curved neck in "La garza": a question mark in the twilight. Still others invoke animals that are notable for their absence, like the firefly in "Luciérnagas," pushed out of the city by light pollution. Through crowdsourcing, Gerber Bicecci adds yet another layer of authorial erasure that deviates from Tablada's constructed immediacy between poet and nature.

Gerber Bicecci's rewriting of Tablada indexes continuities and ruptures in Mexican poetics over the last century. Both use poetry as a site of interplay between word and image; both appropriate prior sources. They diverge in their recognition of this appropriation: Tablada did not openly acknowledge his paraphrasing of the Japanese canon, while Gerber Bicecci encourages readers to intervene upon her text and continue the cycle of revision. As the copyright page notes: "Pueden tacharse, enmendarse o plagiarse creativamente" ([Feel free] to cross things out, revise them, or plagiarize them creatively).[37]

Beyond these formal and methodological distinctions, most notable for our purposes are the ways in which Gerber Bicecci rewrites Tablada by asking how he would have written those poems today. This subjunctive rewriting of *Un día* . . . foregrounds the historical contingency of nonhuman aesthetics, firmly grounded in the observer's temporal-spatial vantage. Reading the two texts side by side, we see this contingent emendation at work as the sublime is dissipated by capital. Tablada's clouds are swift celestial travelers on Andean mountaintops; Gerber Bicecci's are the carcinogenic fumes emitted from mines. Tablada's grapefruit blossom is a delicacy gathered by bees—"huele a cera y a miel" (it smells like wax and honey); in Gerber Bicecci its fragrance is "diluido por el paso / del huracán" (diluted by the passing / of the hurricane).[38] Tablada's palm tree is utterly still in the sleepy afternoon heat; the palm in Gerber Bicecci is "la / extensión monocroma / de las empresas" (the / monochrome extension / of business enterprise). For Tablada, the contemplative appreciation of nature produces a valuable national aesthetic precisely because it escapes the temporal acceleration of modernity and enterprise. By contrast, Gerber Bicecci indicates the inextricability of nature and extractive capitalism. This signals a shift in what each author views as defining a national aesthetic tradition: an eternal essence or the contingent result of modernization. Gerber Bicecci stages this reorientation in her haiku about the bougainvillea. It retrains the eye to see the decorative plant, whose pinks and purples are abundant in tourist imagery of central and southern Mexico, as entangled within the architecture of capitalist modernity, blooming "en / las fisuras de la / maquiladora" (in / the maquiladora's / cracks).

This circumstantial revision of the encounter with the nonhuman reinforces the contingency at the heart of the haiku's poetics. Roland Barthes has called the haiku the "art of the contingent . . . the art of the Encounter." The haiku privileges the particular; it is, according to Bashō (quoted here by Barthes), "simply what happens in a given place, at a given moment."[39] Barthes further observes that in a haiku the poem's subject is not what is most salient. Rather, the referent is "absorbed by the circumstance: what surrounds it, for a lightning-quick moment."[40] It is precisely the circumstantial elements that Tablada stripped from the frame—the sociohistorical context, the material relations that produce the environment—that Gerber Bicecci brings back in. If Tablada were to have written his poems today, Gerber Bicecci suggests, he would not have produced an object that celebrated an unmediated encounter with nature as a moment of individualized transcendental ecstasy. Rather, the emergence of the internet as well as the advent of climate change compel an encounter with the nonhuman as both real and constructed, objectified and agential. For this reason, Gerber Bicecci's haikus relentlessly situate the subject of each poem contiguously: in the shadow of the maquiladora, in the products of industrial monoculture, in the emissions of mines. Betina Keizman draws out this comparison, noting that whereas Tablada's haikus transmit a comforting sense of familiarity and permanence through nature (perhaps in implicit contrast to the turbulence of national history at the time that he wrote *Un día* . . .), in Gerber Bicecci's revisionary take, this certainty is dissolved such that everything is tenuous, in doubt, in decay, but also possible.[41]

Gerber Bicecci's revision of Tablada, combined with the invitation on the copyright page to intervene in the text and make it one's own, positions the text as a commons open to future revision and the creation of future sets. This speculative, unresolved dimension of *Otro día* . . . is reinforced by its visual elements. Gerber Bicecci recreates Tablada's circular watercolor illustrations with photographs altered with acetone, a solvent that she uses to smear or etch shapes into the underlying image. The altered photographs are sourced from the Golden Record that was launched into space in 1977 aboard the Voyagers 1 and 2, probes meant to explore our solar system's outer limits. Conceived as a representative visual and aural archive of humanity, the Golden Record was curated by a committee led by Carl Sagan, who described it as a "hopeful" message in a bottle.[42] Sagan and his team hoped that in the distant future, intelligent life from another planetary system would come across the Voyagers and decipher the Golden Record's time capsule. Such a possibility is remote. Currently, the Voyager probes are drifting in the interstellar medium, the space beyond our solar bubble that exists between star systems. It is estimated that in forty thousand years, Voyager 1 will float near AC +79 3888, a star in the constellation of Camelopardalis, and Voyager 2 will pass near Ross 248, a star in the constellation of Andromeda. If by some chance the spacecraft are found, it will be at a time when humanity most likely no longer exists. The Golden Record is therefore a "posthumous archive" or "eternal homage to a civilization anticipating its own extinction."[43]

Gerber Bicecci notes in the project's artist statement that the acetone interventions on images sourced from the Golden Record stage the erosion of human memory in the far future, when the record is discovered by other life-forms. But as a tool of visual rewriting, the acetone accomplishes more than simply reinforcing the project's speculative thrust. It jeopardizes the integrity of Sagan's visual archive; its haunting, murky presence materializes the histories of damage that the record obscures. The acetone's erasure of the underlying images can be read as a critique of the beliefs that fueled the Voyager expeditions (the techno-utopian desire to transcend the earth,

ENVIRONMENTAL REWRITING 37

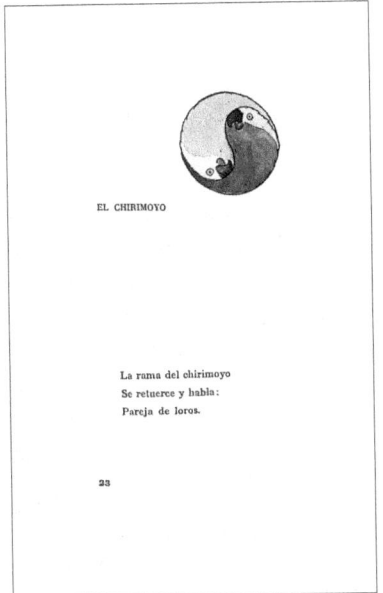

FIGURE 1.1. José Juan Tablada, "El chirimoyo," *Un día . . . (poemas sintéticos)*, 1919. Courtesy of Proyecto José Juan Tablada: Vida, letra e imagen (www.tablada.unam.mx), directed by Rodolfo Mata, Instituto de Investigaciones Filológicas, Universidad Nacional Autónoma de México.

FIGURE 1.2. Verónica Gerber Bicecci, "El chirimoyo," *Otro día . . . (poemas sintéticos)*, 2017. Courtesy of the artist.

the conviction of planetary permanence, a universal account of humanity) as coterminous with processes that have jeopardized planetary life. It also complicates, as Nuria Sánchez notes, the Golden Record's purportedly universal account of humanity: the acetone's distortional presence stands in the place of countless memories excluded by Sagan's team. Sánchez describes this as Gerber Bicecci's "amblyopic" approach to visuality, a method that advocates for a wandering eye, seeing askew through spots and stains, rather than privileging a direct line of uninterrupted sight.[44]

The speculative character of the Golden Record—an archive waiting to be discovered in the deep future after humans have gone extinct—develops the dystopian undertones of the environmental poetics of *Otro día . . .* . In Gerber Bicecci's rewriting of Tablada's ode to the subtropical cherimoya tree, for instance, the cherimoya heralds the emergence of "another tropics" in the polar regions: "Hay otro trópico, / emerge de los polos, / en el deshielo" (There is another tropics, / emerging from the poles, / in the melting ice) (fig. 1.1). This play with the blurred binary between tropics and poles riffs on Tablada's original formulation, structured around a reveal: a swaying cherimoya branch conceals a pair of chattering parrots, the true cause of the branch's motion. Tablada's accompanying illustration features a yin-yang symbol composed of two identical parrot heads (fig. 1.2).

In her revision, Gerber Bicecci replicates Tablada's yin-yang design: a photograph of a field from the Golden Record occupies one side; on the other, a murky cloud of acetone dissolves the image into swirls of sepia. In contrast with the yin-yang's depiction of simultaneous unity and duality, in Gerber Bicecci's rendering, harmonious dualism is unraveled as tropics replace ice. The acetone's revisionary dissolution thus mimics the poem's thematic treatment of planetary warming caused by carbon-intensive economies. Poles and tropics are a binary undone; the cherimoya tree proliferates in new regions, a symbol of regrowth but also imbalance. "El chirimoyo" is a fatalist forecasting of ice melt, but its tone is ambiguous, refusing to equate global warming with the end of the planet or the end of nonhuman life.

Speculative engagement with the far future, perhaps even a future after human extinction, closes *Otro día . . .* in its coda, "Nuevo día" (New day). Tablada's version of this epilogue evokes a dreaming traveler propelled into the future by boat. Gerber Bicecci's re-vision of this future nautical voyager reads as follows: "Testigos de / aplastiglomerado en / el litoral" (Witnesses of / plastiglomerate on / the shore) (fig. 1.3). More so than the Golden Record, the archive left behind by humanity, Gerber Bicecci suggests, is the plastiglomerate planet shaped by consumerist capitalism and its by-products. The material embodiment of capitalism's revisionary geological force, *plastiglomerate* was a term invented in the early twenty-first century to describe stones formed along the shoreline composed of natural debris held together by molten plastic. Plastic waste finds its way to the ocean, where it slowly breaks down into minute particles, entering the food chain and rewriting the shoreline. The timescales of this poem are notably vague, informed by the expansive futurity of plastic, which can take hundreds of years to decompose. Vague too is the subject who witnesses these plastiglomerate shores, who may or may not be human.

Otro día . . .'s back-cover synopsis explains that it conceptualizes the future as "un tiempo que quizás haya dejado de ser nuestro" (a future that is perhaps no longer ours).[45] This subjunctive possibility of a future without humanity, or a future no longer beholden to human authorship, tilts in two directions. It is dystopian in its pessimistic affirmation of capitalist extractivism's destructive teleology, which leads nowhere but to plastiglomerate shorelines. On the other hand, this imagination of a future after the human is not the same as the end of the planet. The continuation of life and poetry

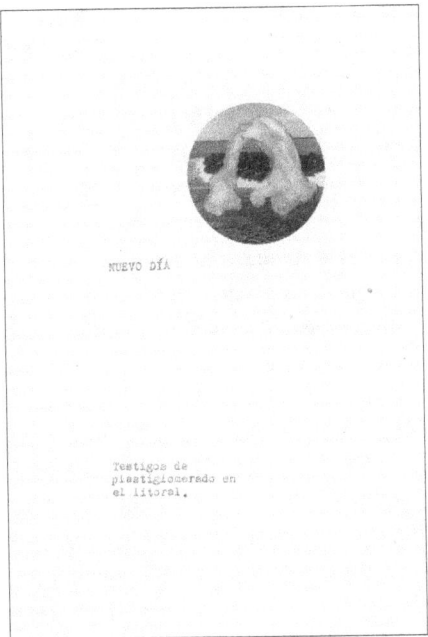

FIGURE 1.3. Verónica Gerber Bicecci, "Nuevo día," *Otro día . . (poemas sintéticos)*, 2017. Courtesy of the artist..

after or beyond the human is condensed in the witnesses of the plastiglomerate shoreline, which go notably undefined, as well as in the haiku dedicated to a spider that "[v]uelve a tejer / después de la explosión / su telaraña" (reweaves / after the explosion / its web). Like the dust that rewrites the desiccated valley in Reyes's "Palinodia del polvo," the concluding poems of *Otro día . . .* position the rewriters of the future as more than human: plastic melding with sand to form new stone, spiders at work in the aftermath of an unnamed anthropogenic disaster.

EXTRACTIVIST SIMULTANEITY IN *LA COMPAÑÍA*

Like *Otro día . . .* 's repurposing of inherited literary form to speak simultaneously to ongoing climate crisis and to a future so distant as to exceed human reach—rewoven, perhaps, by bugs and plastic—*La compañía* also uses rewriting to approximate the temporalities of ecological collapse. But *La compañía* departs from *Otro día . . .* in notable ways, indicating an evolution in Gerber Bicecci's revisionary praxis. Whereas *Otro día . . .* undertakes a globalized exploration of climate crisis, *La compañía* is a localized, site-based interrogation of extractivism in San Felipe Nuevo Mercurio. A silver mining town in Zacatecas that was repurposed as a clandestine waste dump for the toxic by-products of US industry, San Felipe Nuevo Mercurio exemplifies, according to Álvaro García Hernández, "la visión insostenible del desarrollo que aún prevalece en Zacatecas" (the unsustainable view of development that still prevails in Zacatecas).[46]

To tell the story of the town's toxic rewriting by transnational industry, Gerber Bicecci repurposes Amparo Dávila's well-known horror story "El huésped" (The guest, 1959), surfacing the sociohistorical context of extractivism that went uninterrogated in Dávila's work. The inability of Zacatecas's foremost twentieth-century writer to reckon with the true cost of mining and its toxic legacies signals the normalization of instrumentalist approaches to territory at the time, even in feminist texts. Through rewriting, Gerber Bicecci acknowledges her indebtedness to women writers like Dávila, while also reorienting the feminist stance vis-à-vis the environment to one of active ethical engagement. But Gerber Bicecci's engagement with Dávila is not a one-way emendation. Echoing Benjamin's theory of disjunctive percolation, Dávila's incisive diagnosis of the horrors of the patriarchy in "El huésped" structurally informs Gerber Bicecci's take on the ecohorrors and toxic legacies of colonial-capitalist accumulation in *La compañía*, revealing the imbrication of gendered and spatialized power.[47]

Mining in Zacatecas has a long history. The city of Zacatecas was founded in 1546 after substantial silver veins were located by the Spanish in Zacateco territory, prompting a huge influx of settlers.[48] Zacatecas soon became a crucial source of silver for the Spanish Empire, second only to Potosí, Bolivia. Ever since, the fate of the region and of those who inhabit it has been intimately tied to extractivism. Throughout the colonial period, mining fueled regional growth and prosperity. Industry downturns wrecked it. Dips in the seventeenth and mid-eighteenth centuries caused by mercury shortages, mineral deposit declines, and drainage challenges plunged Zacatecas into economic turmoil and catalyzed out-migration.[49]

The industry was reignited during the Porfiriato by the 1892 Mining Law, which offered favorable tax and labor conditions to US companies, displacing artisanal miners.[50] After the revolution, in an effort to reclaim control over Mexico's natural wealth, Article 27 of the 1917 Constitution established that subsoil resources were national property. Nonetheless, the high expense of mining technologies led the revolutionary state to continue to grant concessions to international firms.[51] The global downturn in mineral prices in the early to mid-twentieth century again slowed silver production, leading to the closure of mines and regional depopulation, phenomena registered in the abandoned settings that characterize Dávila's oeuvre, who grew up in the region during that time.⁵²

To reinvigorate the industry yet again, reforms to the Mining Law in the early 1990s established the legal basis for free-market mining and legitimized the expropriation of ejidal and communal lands.⁵³ Concessions ramped up during Felipe Calderón's *sexenio* and continued at a clip during Enrique Peña Nieto's term, incentivized by rising global mineral prices. It is estimated that a third of national lands were leased for mining in concessions granted between 2005 and 2015. The vast majority of these were granted to multinational corporations "at virtually no cost, taking over community territories while polluting their water, land, and air."⁵⁴

Extractivism in Zacatecas is therefore defined by the seemingly infinite reproduction of a boom-bust cycle: a plural and particular apocalypse ongoing since the colonial period. Its expansive temporality has yielded a similarly expansive material legacy: from the minute toxins that persist in groundwater centuries later to the massive rewriting of landscapes through the removal of soil. Rocio Gomez stresses that extractivism touches nearly every aspect of regional life, so much so that "Zacatecas and silver mining [are] biologically intertwined and indistinguishable."⁵⁵ Its formative force is evidenced by deforestation and desertification, as well as by countless material traces, such as the "network of dams, reservoirs, and millraces for washing ores" and "piles of tailings, mine waste, and scoria (pebble-sized twists of furnace discards), all heavily mineralized, open to the elements" that dot the Mexican mining belt.⁵⁶ Processes like smelting, chemical leaching, and mercury amalgamation, integral to the transformation of ore into silver, leave chemical compounds in their wake. These substances make their way into waterways and aquifers, where the Comisión Nacional del Agua (CONAGUA, National Water Commission) has found abnormally high levels of lead and mercury, meaning that even those who do not participate in the industry cannot escape its effects.⁵⁷ The disastrous consequences of mining on environmental and human health have been contested by countless communities, making mining the biggest driver of socioenvironmental conflict in Mexico.⁵⁸

Mining in Zacatecas can be understood as an imposed process of cyclical territorial rewriting buoyed by the ever-deferred promise of regional prosperity, a revisionary process that Gerber Bicecci's intensive method of rewriting emulates. *La compañía* was commissioned for the XIII FEMSA Biennial hosted in Zacatecas in 2018 and curated by Willy Kautz around the theme "Nunca fuimos contemporáneos" (We were never contemporary), an allusion to Bruno Latour's *We Have Never Been Modern*. There Latour

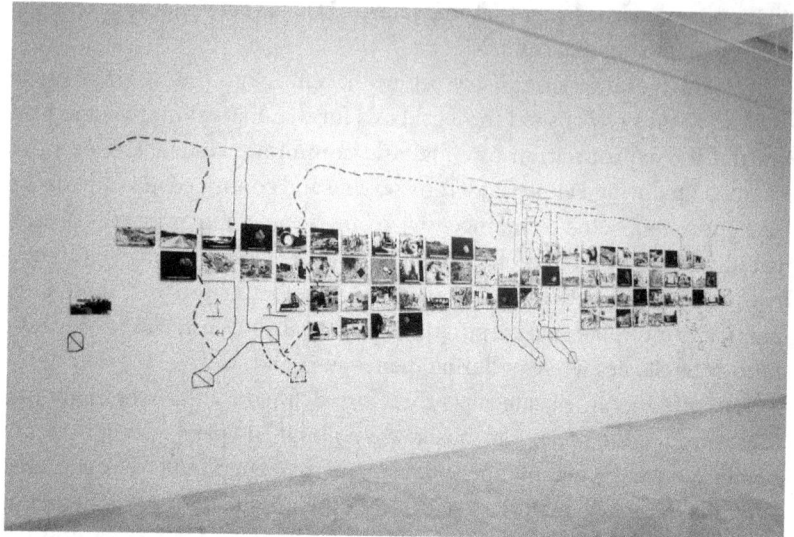

FIGURE 1.4. Installation view of Part A of Verónica Gerber Bicecci's *La compañía* at the XIII FEMSA Biennial in the Museo de Arte Abstracto Manuel Felguérez, Zacatecas. Courtesy of the artist.

argues that the so-called modern era is not so much a rupture from but the continuation of the colonial quest for mastery over life—a conceit that also underlies *La compañía*.[59] The XIII FEMSA Biennial invited artists to create works in dialogue with Latour or with the host state of Zacatecas. Many, including Gerber Bicecci and performance artist Naomi Rincón Gallardo, whose work I discuss in the next chapter, chose to engage with both by addressing Zacatecas's ties to mining. Registering the simultaneity of extractivism's past, present, and future effects, *La compañía* configures mining in Zacatecas as an ongoing apocalypse, or, as Gerber Bicecci puts it in the title of the accompanying digital media project, "a dystopian machine."[60]

Gerber Bicecci's contribution to the XIII FEMSA Biennial was made up of three parts, *La máquina distópica*, an interactive web oracle codesigned with Canek Zapata and Carlos Bergen, and two other components, Part A and Part B, which were subsequently integrated into the bound iteration of *La compañía* published by Almadía.[61] Part A of *La compañía*, which I turn to in greater detail ahead, rewrites Dávila's short story "El huésped" into a prophetic tale of extractivist horror. The second part of *La compañía*, Part B, assembles a collective account of San Felipe Nuevo Mercurio by interweaving a constellation of sources: ethnographic snippets of Gerber Bicecci's conversations with local miners and scholars, excerpts from scientific papers, geological surveys, mine blueprints, environmental reports, diplomatic

FIGURE 1.5. Installation view of Part B of Verónica Gerber Bicecci's *La compañía* at the XIII FEMSA Biennial in the Museo de Arte Abstracto Manuel Felguérez, Zacatecas. Courtesy of the artist.

communications, and so on. In the FEMSA debut of *La compañía*, Part A was mounted to the wall of the Museo de Arte Abstracto Manuel Felguérez in the format of a subterranean mine map, allowing visitors to read it piece by piece or observe it as a whole (fig. 1.4). By contrast, Part B was displayed as a stack of black paper on the floor in the center of the exhibit space (fig. 1.5). Biennial visitors who wished to engage with this portion of the project could do so only by crouching on the floor, as if excavating an archive from the earth.

By assembling a chorus of interwoven voices, Part B suggests that the story of San Felipe Nuevo Mercurio cannot be told from any singular vantage. The interwoven snippets echo the cinematic technique of montage, the intercutting of fragments to expeditiously gesture toward a whole. This method situates extractivism as relationally produced: shaped by meteor showers and surveyors, registered by journalists and official documents, made legible and legitimized by scientific study, and contested by those who experienced its harmful effects. In tacit recognition of her outsider status in Zacatecas, this compositional method positions Gerber Bicecci (who is based in Mexico City) more as an assembler than author, whose work lies in seeking out, listening to, selecting, and arranging information. As Alicia Sandoval puts it, "aunque Gerber Bicecci no redacta, sí construye un relato" (although Gerber Biccecci doesn't write, she does construct an account).⁶² She shapes the narrative by choosing what to include and what not to include, a method that, like rewriting, downplays authorial voice but not necessarily narrative control.

Similarly informed by a method that strains against the bounded individualism of authorship, Part A of Gerber Bicecci's *La compañía* rewrites Amparo Dávila's short story "El huésped." Dávila is now recognized as one of Mexico's foremost authors of fantastical fiction. But when she published her work in the mid-twentieth century, it gained little traction. In large part this is because she was a woman writing in a sexist industry that systematically aligned assessments of genius with masculinity.⁶³ This gendered marginalization was compounded by Dávila's distance from dominant literary trends "rooted in strong allegorizations of the nation."⁶⁴ Dávila's stories instead dug into rural life, domestic spaces, and womanhood. Through genres like the fantastic and horror, she explored the underbelly of the human psyche and states of fear, solitude, death, and madness.⁶⁵

In the twenty-first century, Dávila has risen to broad acclaim thanks to the efforts of women authors who identify her as an important antecedent. Cristina Rivera Garza was the first to do so, in *La cresta de Ilión* (The Iliac Crest, 2002), followed by Guadalupe Nettel, who riffed on Dávila's "El huésped" in an eponymous novel published in 2006. These recuperative efforts brought Dávila's work to national and international attention, leading to her consecration by the Fondo de Cultura Económica in 2009 and the translation of her work into English for the first time in 2018 by New Directions, just two years before her death.⁶⁶

By rewriting "El huésped," Gerber Bicecci joins her peers who use rewriting to position themselves within a feminist lineage of women's writing in Mexico. What differentiates her rewriting from the work of her contemporaries is its formal intensity. Gerber Bicecci replicates Dávila's best-known story almost verbatim, modifying only a select set of properties: verb tense, narrative point of view, and the names of two characters. This revisionary project is even more intensive than her engagement with Tablada, whose textual infrastructure she adopts while liberally modifying its content. Whereas Rivera Garza and Nettel take up Dávila to explore the relationship between writing practices and gender, Gerber Bicecci rewrites Dávila in order to reorient contemporary interpretations of her work back toward its space of production: the extractive zone of Zacatecas. In so doing, Gerber Bicecci draws connections between gendered violence and the violence of extraction, signaling them as coconstitutive. Her work also signals a broader shift in feminist writing toward greater ethical engagement with territory; this growing ecofeminist corpus also includes Rivera Garza's latest works as well as the books of poetry by Isabel Zapata, Karen Villeda, Xitlálitl Rodríguez Mendoza, and Maricela Guerrero examined in Chapter 3.

Dávila grew up in Pinos, Zacatecas, a small mining village that declined along with the mining industry after the revolution. As silver slumped, Pinos emptied out. It was a ghost town, Dávila wrote of her cloudy, mountainous hometown, as desolate as Juan Rulfo's Luvina:

> Pinos es un viejo y frío pueblo minero de Zacatecas con un pasado de oro y plata y un presente de ruina y desolación. Yo nací en la casa grande del pueblo y a través de los cristales de las ventanas miraba pasar la vida, es decir la muerte, porque la vida se había detenido hacía mucho tiempo en ese pueblo.

> Pinos is an old, cold mining town in Zacatecas with a past of gold and silver and a present of ruin and desolation. I was born in the biggest house in town and through the windowpanes I watched life go by, which is to say I watched death, because life had been stopped for a long time in that town.[67]

This moody atmosphere, Erica Frogman-Smith posits, "closely resembles the often fantastic atmosphere" of Dávila's fiction.[68]

Given the centrality of mining to the places where Dávila grew up—Zacatecas and San Luis Potosí (where she moved at the age of seven)—it is curious how rarely it is mentioned in her work. For Dávila the problem was not so much the industry's presence but its absence. At the time she wrote, mining had entered a period of significant decline, resulting in regional depopulation, registered in the abandoned and run-down spaces that characterize her narrative. In Dávila's poem "Espejo lento" (Slow Mirror), the speaker finds herself left behind, alone, "como plata fría, / como tierra en olvido, vacía" (like cold silver, / like forgotten land, empty).[69] Implicit in the alliance between the lonely speaker and the silver forgotten in the earth is a nostalgia for the extractive boom, without which land, mineral, and poet are devalued. Dávila's affirmative view of mining reflects the contemporaneous developmentalist faith in extractivism as the region's reason for being, instantiated since the colonial silver boom.

Neither mining nor Zacatecas are explicitly mentioned in "El huésped," but the stifling atmosphere endured by its protagonist resonates with Dávila's experience of a town left behind by industry. "El huésped" narrates the experience of an isolated upper-class housewife who lives in a sprawling home with her children and maid. She is trapped in an unhappy marriage with a man who is infrequently home and treats her dismissively, as if she were just another piece of furniture. One day the husband brings home a guest. The guest—left deliberately undefined in the text—has monstrous feline qualities. It lurks around the house, snacking on raw meat, its yellow eyes fixed on the protagonist and her children. Its menacing presence contaminates the domestic space, transforming it from a refuge into a claustrophobic nightmare. Yet the protagonist's husband refuses to believe that the guest is a threat, minimizing her concerns as overblown. This oppressive atmosphere is compounded by the narrator's isolation. With no friends, extended family, or resources, she is dependent on her husband's largesse. This precarious situation escalates when the guest attacks the maid's child. Forced to defend themselves, the two women band together and trap the guest in a room while the husband is out of town. There they contain it for several days, until it dies from starvation.

"El huésped" is a masterful exploration of the horror of female existence in the patriarchy. While some critics have discussed the productive ambiguity of whether the guest is real or a figment of the protagonist's imagination, most agree that the guest is an allegorical manifestation of domestic violence: a menacing presence the husband insists is not a problem.[70] Dávila exploits the conventional tropes of the horror genre—isolation, confinement, paranoia, helplessness, a threatening Other—to give shape to the claustrophobic terror experienced by abused women. The protagonist is trapped, with no financial autonomy or external support; her children are similarly vulnerable. Nonetheless, the story's conclusion formulates a message of female agency. Together the women kill the guest by isolating and enclosing it. This vengeance is only possible thanks to the women's cooperation, an alliance across class (and, implicitly, race) that articulates the power of an expansive feminism.

Gerber Bicecci replicates Dávila's story word for word, enacting a sort of dwelling in the inherited text. Within these imposed constraints, Gerber Bicecci performs four interventions. First, she systematically changes Dávila's use of the present tense to the future tense. Second, she alters the narrative point of view from first person to second person. Gerber Bicecci's grammatical interventions augment the story's sense of dread, foretelling a future harm that will be enacted on the reader. What has happened to Dávila's protagonist, Gerber Bicecci affirms, will happen to you. Third, Gerber Bicecci alters the identities of two main characters. In her retelling, the sinister guest becomes the Company, and the maid becomes a machine.

The fourth intervention is the reterritorialization of Dávila's text to San Felipe Nuevo Mercurio through dialogic interplay with image. Graphically, each sentence is presented on a separate page, set off against black-and-white photographs of the Nuevo Mercurio mine taken by Elizabeth del Angel (studio photographs of ore), Dr. Héctor René Vega-Carrillo (archival

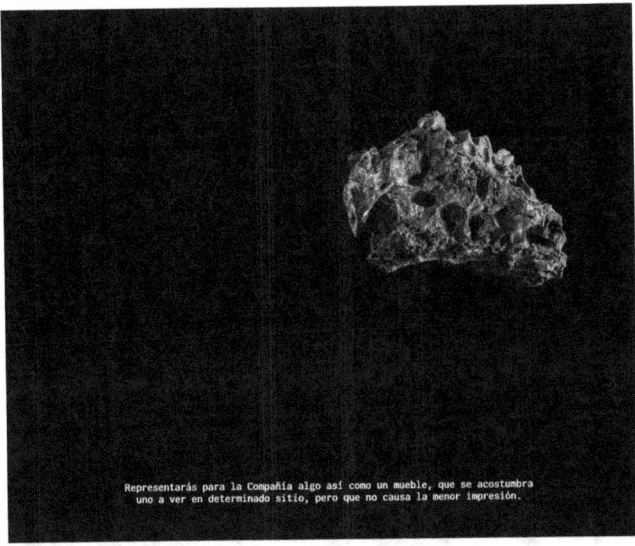

FIGURE 1.6. Dialectic interplay between Amparo Dávila's short story "El huésped" and studio photographs of ore by Elizabeth del Angel calls attention to the naturalization of ore as a commodity. Verónica Gerber Bicecci, *La compañía*, 2019. Courtesy of the artist.

FIGURE 1.7. Part A of Verónica Gerber Bicecci's *La compañía* reterritorializes "El huésped" by overlaying it on photographs of the abandoned San Felipe Nuevo Mercurio mine. *La compañía* simultaneously dwells in the territory of San Felipe Nuevo Mercurio and in Amparo Dávila's textual infrastructure. Verónica Gerber Bicecci, *La compañía*, 2019. Courtesy of the artist.

FIGURE 1.8. Verónica Gerber Bicecci, *La compañía*, 2019. Courtesy of the artist.

photographs from 1986), and Gerber Bicecci. Some images are superimposed with fragments appropriated from Manuel Felguérez's abstract work *La máquina estética* (1975), which Gerber Bicecci uses to represent the maid/machine; Felguérez is also from Zacatecas. The black-and-white images enhance the foreboding ambiance of Dávila's story and ground it in the material specificity of San Felipe Nuevo Mercurio as the site of Dávila's claustrophobic nightmare (figs. 1.6 and 1.7).

With these modifications, Gerber Bicecci transforms Dávila's horror story about a woman trapped in a situation of domestic abuse into one about life in the "extractive zone," defined by Macarena Gómez-Barris as organized by the colonial paradigm of "capitalist resource conversion."[71] Like the emotionally manipulative husband who insists that the guest is not a threat in spite of all evidence to the contrary, private and public forces behind mining dismiss community concerns about its harmful effects. In the context of the contemporary "commodity consensus," in which high mineral prices have led Latin American governments across the political spectrum to embrace extractivism, critics of extractivist projects (who are often Indigenous and/or rural poor) are discredited by political elites as holding "irrational" or "anti-modernity" beliefs or as being the naive puppets of foreign NGOs.[72] This reinforces extractivism as the only horizon of political possibility, just as the husband in "El huésped" indifferently insists that "La Compañía es completamente inofensiva [. . .] te acostumbrarás" (The Company is totally harmless . . . you'll get used to it) (fig. 1.8).[73]

The titular protagonist of Gerber Bicecci's *La compañía* alludes to Minera Rosicler, a US corporation that operated the Nuevo Mercurio mine from 1940 to 1970 to extract cinnabar, the source of mercury. At its peak, the population of the surrounding community of San Felipe Nuevo Mercurio rose to 10,000. In the 1970s, scientific studies linked mercury to an array of environmental and health problems, leading to greater regulation and a subsequent drop in price. Reduced demand combined with a dwindling deposit led Rosicler to cease operations. Shortly thereafter, the owner, John Nugent, and his son repurposed the abandoned mine into a clandestine waste dump. From 1972 to 1979, the Nugents transported illicit shipments of hazardous substances from Houston, Texas, to Nuevo Mercurio.[74] The waste dumped in Nuevo Mercurio was generated by companies like Diamond Shamrock, Monochem, B. F. Goodrich, PPG Industries, and Monsanto. Investigators found that Rosicler imported more than 4,500 tons of hazardous waste in the 1970s, including 635 200-liter metal storage drums of toxic waste. Forty-two of these drums contained polychlorinated biphenyl (PCB), a highly carcinogenic manmade organic chemical used in heat transfer fluids. This is likely a conservative estimate, given that some barrels had been emptied and repurposed to hold water by villagers unaware of their contents.[75] This unsettling fact dramatizes the terrifying disjuncture of inviting something into one's home that is actually an existential threat. Rosicler was invited into Nuevo Mercurio by the patriarchal state, where it engaged in monstrous behavior that was initially invisible and thus easy to deny but that ultimately devastated the community.

The dumping ground for the unwanted toxic by-products of multinational corporations, San Felipe Nuevo Mercurio embodies the structural and spatialized inequalities of extractivism. The toxins deposited there rewrote its future: today San Felipe Nuevo Mercurio is the greatest site of PCB contamination in all of Mexico.[76] Exposure to PCBs has been associated with skin rashes, numbness, neurodevelopmental disorders, and decreased cognitive functioning. A team of scientists assessing soil in the town in 2011 also found extensive mercury contamination, as well as dangerous levels of both PCB and mercury in blood tested from local children.[77] The Mexican government attempted to litigate the case against Rosicler in the early 1980s but was stymied by ambiguous international environmental laws. Even though the state had been aware of the extensive contamination in San Felipe Nuevo Mercurio since the 1980s, it was only in 2002 that it designated resources for remediation. Nothing happened until 2011, when the government finally removed eighty barrels of toxins that had been left behind and 750 cubic meters of contaminated earth. Without safe access to land and water, San Felipe Nuevo Mercurio is now populated by only a few hundred people. It materializes extractivism's horrific spatial and temporal simultaneities, in which the future is endlessly enfolded into past harm. In this sense, Gerber Bicecci's rewriting of Dávila parallels Rosicler's toxic rewriting of Nuevo Mercurio: a contained space that has been comprehensively transformed.[78]

The locus of fear in Gerber Bicecci's rewriting is the Company, rather than toxins themselves, or the disabled landscapes or bodies they engender.[79] The Company in Gerber Bicecci's rewriting manifests the threat of contamination, which, if we remember, first attacks the maid's children. Part of what makes Dávila's guest so terrifying is its ontological ambiguity; to the reader, it never comes fully into view. Toxic waste is similarly hard to pin down. Beyond the emblematic metal drums, toxicity goes unseen, visible only in its material effects. Chemicals like PCBs are notably persistent. They remain in the soil, air, and water without degrading for decades or even centuries and are given new life in crops and livestock, "extending racism in time," as Michelle Murphy writes.[80] A line in *La compañía* underscores this enduring threat: "Aunque la Compañía pasará todo el día durmiendo, no podrás confiarte. Habrá muchas veces que, cuando estés preparando la comida, verás de pronto su sombra proyectándose sobre la estufa de leña." (Even when the Company will spend all day sleeping, you won't be able to trust it. There will be many times, when you are preparing food, that you will suddenly see its shadow looming over the woodstove.)[81] Even in Rosicler's absence, its toxic wake looms over quotidian routines, conditioning all bodies in spatial and temporal contiguity, effectively collapsing public and private space, as Gerber Bicecci indicates by using the house in "El huésped" to stand in for the territory of San Felipe Nuevo Mercurio. Like the guest, the Company slips past institutional accountability, exploiting legal loopholes and the general disposition of the state to maintain unconditional rapport with investors. And like the dismissive husband, the state is complicit in the Company's violence in its silencing of those who raise concerns.

Gerber Bicecci's use of the prophetic future tense in her rewriting predicts the industry's future continuation as inevitable. History suggests this is the case. In his analysis of mining in Mexico, Daviken Studnicki-Gizbert has shown that little has changed since the colonial era. Five hundred years ago there were approximately 450 active mining districts in New Spain. After many iterations of the boom-bust cycle, that number remained surprisingly steady at the turn of the twentieth century, but at that time, they were managed by companies from the United States. A century later, as of 2011, Mexico is host to nearly seven hundred mines, in the same districts mined since the colonial era, but these days run primarily by Canadian transnationals.⁸² Through new technologies or new laws, capitalism finds a way to push past the limits of exhaustion. We see this at work in the fact that within a single decade, from 2000 to 2010, transnational mining companies in Mexico extracted double the entire amount of gold unearthed by the Spanish in three hundred years of colonial rule.⁸³ We also see this at work in Rosicler's repurposing of the abandoned mine into a transnational toxic waste dump.

As a diagnostic that speaks to the history of extractivism in Mexico as well as to its ongoing intensification, Gerber Bicecci's use of the future tense to predict the continuation of extractivist harm is a potent framing device that emulates extractivism's grammar of simultaneity and totalizing temporal reach. But it also runs the risk of pathologizing *La compañía*'s subject, San Felipe de Nuevo Mercurio, circumscribing this space as doomed to no other possible future. Along these lines, Eve Tuck has powerfully critiqued "damage-centered research," cautioning that even when such work is performed with antiracist intentions, it can nonetheless reinforce the depiction of marginalized communities as "defeated and broken," singularly defined by oppression.⁸⁴ (Tuck advocates for research that centers desire rather than damage, an approach adopted by the artists and activists considered in the next chapter.) It is worth considering whether *La compañía* falls into this trap. The black-and-white images of San Felipe Nuevo Mercurio in Part A include no human subjects, a decision that generatively resists framing the human body as the locus of horror but also erases the families who remain and continue to desire and advocate for the community's future.⁸⁵ It abstracts San Felipe Nuevo Mercurio into an allegory of extractivist horror, which reinforces its conceptual portability but erodes its lived specificity. This abstraction is tempered by the chorus of voices collated in Part B, which offers a different narrative entry point to the territory through montage.

But while Gerber Bicecci's use of the future tense might seem to affirm Zacatecas's foreclosure, recall that Dávila's tale provides a blueprint of response: a way to defeat the sinister guest. *La compañía* tells the reader of the fear they will face, but also of the action they will take, following in the steps of Dávila's narrator, the victim of domestic abuse who smothers and starves out the guest with the help of her maid, an allegory for political allyship across lines of class and race. This is no easy win: "Los días que seguirán serán espantosos. La Compañía vivirá mucho tiempo sin aire y sin alimento. Al principio golpeará la puerta, se tirará contra ella, gritará desesperada, arañará." (The days that will follow will be terrifying. The Company will live for a long time without air and food. At first it will bang on the door, throw itself against it, it will desperately shout, clawing at the door.)⁸⁶ The prophecy of

FIGURE 1.9. Verónica Gerber Bicecci, *La compañía*, 2019. Courtesy of the artist.

continued extractivist harm is thus also a rallying cry. The Company, like the guest, threatens all sectors of society—albeit unevenly—requiring alliances across class and race to defeat it.[87]

La compañía cultivates these allyships through the use of the second person, directly embroiling readers in the anticipation of the expansion of what Maristella Svampa calls the "extractive frontier."[88] The shift from Dávila's omniscient narrator to second person suggests that Gerber Bicecci perceives distance as a primary obstacle in making extractivism legible as a shared political concern; her project attempts to redress spatial and psychological distance from the extractive zone through direct address. *La compañía* foretells readers of their future interaction with extractivist enterprise and tells them how it will feel: "No podrás reprimir un grito de horror cuando veas a la Compañía por primera vez. Será lúgubre, siniestra." (You won't be able to repress a shout of terror when you see the Company for the first time. It will be lugubrious and sinister) (fig. 1.9).[89] This anticipatory subjunctive interpellation draws from the affective strategies of horror, using the text as a point of contact with the reader's body. *La compañía* does not just depict the horror of past damage, but demands that readers imagine themselves entangled in that landscape, "translating [the] sensorial enactments [of horror] across our bodies."[90]

This anticipatory address presumes the presence of readers who do not already live in the extractive zone and thus need to imagine what that harm would be like. It aims to erode the physical and emotional distance that many consumers of high art—like the attendees of the FEMSA Biennial in Zacatecas, scholars like me, Almadía readers, and even Gerber Bicecci herself, who lives and works in Mexico City—have with the environmental and social costs of the energy and production systems that fuel their lives. The invisibility of these harms is engineered by private/public extractive industries to diminish opposition. Class privilege and spatial distance from the toxic effects of extractivism make them less urgently felt, particularly as reckoning with extractivist systems means questioning the economic and infrastructural underpinnings of modern life. As Elizabeth Barrios has suggested, drawing on Robert Johnson's concept of the "fossil unconscious," "something on a psychic level leads us to resist grappling with fossil fuel energy," because it destabilizes the basis of modern life as we know it.[91] These observations can be extended to mining, particularly in Zacatecas, where it has defined the region's economic and cultural identity since the colonial period. The efficacy of this psychic distancing is evidenced by Dávila's own silence on the subject.

Through rewriting Dávila's "El huésped," Gerber Bicecci produces a prophetic account of ecohorror that predicts the continuation of extractivist harm in Zacatecas, but also its eventual eradication through collective action along the lines sketched out by Dávila's tale.[92] If extractivism is a form of planetary rewriting, a destructive form of world building powered by ever-deferred promises, rewriting as an artistic method runs countercurrent to this foreclosure by its methodological insistence that any given text—or ecological space or economic system—always remains open to future revision. In an interview, Gerber Bicecci articulated the importance of revisionary praxis:

> Tenemos que cambiar nuestra estructura de pensamiento, porque es la única manera de encontrar un futuro posible. Al final, hablar de la crisis climática y de las ruinas de una mina es tratar de imaginar y recuperar nuestras posibilidades de que haya un futuro para todos. Creo que si seguimos viendo y pensando el mundo como hasta ahora lo hemos hecho no vamos a encontrar salida.

> We must change our way of thinking, because it is the only way to find a possible future. Ultimately, talking about the climate crisis and an abandoned mine is a way of trying to imagine and recuperate the possibility of a future for everyone. I think that if we continue to see and think about the world as we have up until now, we won't find a way out.[93]

With this quote, Gerber Bicecci makes a bid for art's subjunctive orientation that echoes the position laid out by Amitav Ghosh in the Introduction, where he argued that to "think about the world only as it is amounts to a formula for collective suicide." Rewriting, for Gerber Bicecci, is an artistic practice that actualizes fiction's ability to imagine other possibilities. As Graciela Speranza observes, *La compañía* indexes the power of art to put forth new forms out of the ruins of history.[94] Her revisionary praxis reckons with subordination to established historical and material circumstances, while simultaneously making moves to amend the given, a tension at the core of subjunctive aesthetics in the era of climate change.

REWRITING IN/AS THE ANTHROPOCENE

What revisionary modes does the Anthropocene activate or require? In "Palinodia del polvo," Alfonso Reyes lamented the aesthetic loss caused by centuries of anthropogenic desiccation, backpedaling his endorsement of the very project that he had earlier posited formed the basis of national identity. As the colonial rewriting of Anáhuac doubled back, exceeding and confounding expectations, Reyes found the outcome grotesque. Dust storms now wrote the sensorial experience of the valley, rendering it a monotonous hue. This material transmutation in turn revised the meaning of Reyes's iconic phrase, "la región más transparente del aire." With its origins in Humboldt's gaze and its future yoked to Mexico City's smog, Reyes's slogan reveals the Anthropocene as at once the result of revision and its driver, a revisionary force that transforms the meaning of language itself.

Even though Reyes could not have known what the future would bring, reading his *Visión* subjunctively, as if it had anticipated present-day air contamination in Mexico City caused by the confluence of desiccation, industrialization, and car culture, allows us to see environmental crisis as part and parcel of the national project. This subjunctive engagement with the literary past underlies Verónica Gerber Bicecci's environmental rewriting of José Juan Tablada and Amparo Dávila, an exercise whose point of departure is reimagining a text as if it were written today. Following Benjamin, such a method generatively "puts the truth of all present action to the test" and "serves to ignite the explosive materials that are latent in what has been."[95] Yet the temporal affordances of the "as if" also point to a third and future set, which Gerber Bicecci invokes through the future tense in *La compañía* and the invitation to readers of *Otro día*... to intervene in her text and remake it. This simultaneous straddling of past, present, and future differentiates Gerber Bicecci's rewriting from other recent iterations of environmental rewriting.

In addition to using rewriting to reassess instrumentalist approaches to the nonhuman environment, Gerber Bicecci's method demonstrates how literature and art can approximate the temporal demands of the climate crisis, which requires that we at once project forward to anticipate its future evolution, look back to its origins, and reconsider its present normalization. The "as if," a mode of thinking proper to the subjunctive, rises to meet these demands, zigzagging among timescales to materialize the simultaneity of the Anthropocene.

So what does the advent of environmental rewriting as method tell us about authorship in the era of climate change? In "The Death of the Author," Roland Barthes famously argued that the meaning of any given text is ultimately determined, not by the author, but by the text's present and future readers.[96] What Barthes described was the relational production of meaning over time as a text is transformed through its reception, a process rewriting makes explicit. Thinking with Barthes from the vantage of the Anthropocene adds yet another twist, acknowledging that meaning is also shaped by nonhuman agents and planetary conditions. For Reyes, the radical decentering of human authorship over the valley's aesthetics, and the concomitant loss of control over the meaning of his iconic phrase, was horrific. If Reyes located the locus of horror in the agency of dust, for Gerber Bicecci, nonhuman participation in the process of anthropogenic planetary rewriting is simply a sign of the inherent relationality of ecological and artistic creation. This human and nonhuman reweaving is ambiguous: it can lead to centuries of toxicity or to the creation of new images produced by acetone's interaction with photographs.

This leads us to rethink what it means to write in the Anthropocene. Counter to current market imperatives that stress individual agency, originality, and novelty, environmental rewriting positions cultural production as fundamentally relational. In Gerber Bicecci's hands, the work of art comes into view as an ecosystem: composed over time by a variety of actors, centrifugal forces, and sedimented layers. The existing forms of Tablada's and Dávila's texts that Gerber Bicecci works within constrain her agency. She moves within these inherited and extant forms, a method that positions the author as a creator who operates within a diffuse web. And yet within these established forms, there is room for contingency, a way to write in chance and change.

The centrality of collaboration to contemporary rewriting has been theorized by Cristina Rivera Garza under the term "disappropriation." In implicit contrast to appropriative forms that extractively mine the work of others for personal gain, disappropriative writing "explodes the plurality that precedes individuality in the creative process, opening a window onto the material layering so often concealed by appropriative texts."[97] Rivera Garza describes this material layering as "geological," in a nod toward how language is constructed over time through the processual circulation of ideas among writers and publics, as well as how writing is part of an ecosystem sustained by a multitude of actors, involving editors, curators, delivery people, printers, bookshop clerks, janitors, and librarians, funding entities like the state, galleries, and FEMSA, and the material world, the energy and minerals that sustain computer hard drives and the trees pulped into paper. Gerber Bicecci's environmental rewriting shares many of the precepts outlined by Rivera Garza, including the focus on material layering, collaborative work, and future orientation. The apocalyptic inclinations of her rewriting—the prediction of extractivism's continuation, the gesture to a future after human extinction—acknowledge that writing and cultural production exist in a world structured by extractive capitalism. But rewriting proffers a subjunctive method for processing anxiety about the world through small lateral movements of repetition and re-creation that mobilize the "as if." Through rewriting, Gerber Bicecci demonstrates the possibility of reformulating our approach to nonhuman life from one of ownership and authorship to one of collective creation.

CHAPTER 2

Land Defense and Counterfactual Mourning

In February 2019, Nahua land defender and radio broadcaster Samir Flores Soberanes was murdered in his home in Amilcingo after months of vocal opposition to Proyecto Integral Morelos, a planned megadevelopment project that would construct a thermoelectric plant and a natural gas pipeline crisscrossing Indigenous and ejido lands in Morelos. As an organizer for the Frente de Pueblos en Defensa de la Tierra y el Agua de Morelos, Puebla y Tlaxcala (FPDTA-MPT, Peoples' Front in Defense of the Land and Water of Morelos, Puebla, and Tlaxcala) and the Consejo Indígena del Gobierno (CIG, Indigenous Governing Council, a Zapatista-related network of Indigenous groups), Flores gave voice to community objections to the megaproject, including the contamination risk the pipeline poses to the water supply and the vast quantities of water (fifty million liters a day) that will be redirected from the Cuautla River to cool the thermoelectric plant turbines, resulting in water scarcity for thirty surrounding ejidos.

Fellow activists in FPDTA and CIG contend that Flores's murder was an attempt to silence dissent to Proyecto Integral Morelos, pointing to the fact that it occurred just days before a planned referendum of the megaproject. After Flores's murder, activists called for the consultation vote to be postponed. Dismissing these objections, authorities went forward with the referendum, which passed with 60 percent support, with many of the most affected communities refusing to participate in a process they saw as invalid and coerced.[1] Instead, in the town of Huexca, the planned site for the thermoelectric plant, citizens came together as an *asamblea popular* (people's assembly) to gather signatures to signal their collective rejection of the project.[2]

Unlike the accretive violence discussed in Chapter 1, the violence experienced by land defenders in Mexico is better understood as "violencia desnuda" (bare violence), defined by Marxist sociologist Jaime Osorio as "una violencia obscena, con un exceso de realidad, que a fuerza de repetirse uno y otro día va perdiendo su capacidad de horrorizar" (an obscene violence, excessively real, whose capacity to create horror is dampened by its daily repetition).[3] Mexico is one of the deadliest countries in the world for land defenders. Between 2012 and 2021, at least 154 land defenders were murdered and hundreds more intimidated. This pervasive violence, according to Osorio, demonstrates that consent to extractive projects in Latin America is a fiction, since capital forcibly obligates those in its path to put themselves at its disposition.[4] Violence against land defenders is overt, swift, brutal, and normalized by its recurrence—yet nonetheless largely invisible in the mainstream media.

In the wake of Flores's murder, Indigenous and environmental activists honored his life and called attention to his death through the slogan "Samir vive, la lucha sigue" (Samir lives, the fight continues), spreading this rallying cry on social media and inscribing it on walls in cities throughout Mexico.[5] Borrowed from the Zapatista formulation "Zapata vive, la lucha sigue, sigue," the slogan in this context denies murder its conclusive force. Even though the phrase "Samir vive" does not grammatically encode the subjunctive, its counterfactual force situates it within the realm of subjunctive aesthetics. It suggests this death need not have been so, it could have been otherwise. Its negation of Flores's death contests the foreclosure of political dissent to extractivism by affirming the continued existence of other possible worlds: a world where Samir Flores lives and in which relations with territory are forged around consensus, relationality, and life.

What does it mean to mourn in the subjunctive? Mourning is often understood as the attempt to confront, process, and overcome past trauma. Idelber Avelar describes mourning in the context of the postdictatorial Southern Cone as an "active forgetting" that runs counter to "passive forgetting," the key operation by which the neoliberal state represses "its barbaric origins" in authoritarian violence and relegates "the past to obsolescence... because the market demands that the new replace the old without leaving a remainder."[6] Applied to the context of land disputes in Mexico, "passive forgetting" aptly portrays how the state and the media downplay the foundational role of bare violence and dispossession in the extractive economy. Protests, memorials, graffiti, and murals that mourn murdered land defenders in the public sphere perforate passive forgetting.

Yet "active forgetting" does not quite describe the intent behind counterfactual mourning, since its goal is not to "come to terms with a past catastrophe" and overcome it, nor is its engagement with loss melancholic.[7] This is because violence against land defenders is a trauma that is not confined to the past, but rather is ongoing and constitutive of extractivist modernity. Consequently, counterfactual practices of mourning are not actually oriented toward the state or the justice system, as is the case in the postdictatorial context, or even in other contexts of mourning in Mexico, like the Tlatelolco or Ayotzinapa massacres, which are containable and thus redressable as events. The purpose of counterfactual mourning exceeds what the state can or is willing to do and exceeds the scope of any individual case that might be brought to trial. Rather, the work performed by counterfactual mourning is imaginative and future oriented in its utopic invitation to collectively join in the subjunctive act of enacting other possible worlds that might exist beyond extractivism and its corollaries of violence and dispossession, worlds where the people engage with territory on the basis of life, community, and desire.

Visual and discursive acts of counterfactual mourning refer to death but deny it as such, rerouting back to life in a subjunctive expression of desire for how the world *could have been* or *could still be*: "Samir vive." Counterfactual mourning embraces the hauntological, propelling the dead into the future through the continued affirmation of their life and the futurity of land defense.[8] It affords the victim space in the collective body-territory, it asserts their vitality, and it acknowledges the public's responsibility to them and the political projects they died defending. Akin to an imaginative act of "undoing," counterfactual mourning produces what Catherine Gallagher describes as "an enlarged sense of temporal possibility correlating with a newly activist, even interventionist, relation to our collective past."[9] Just as ghosts disturb the binary separation of the living from the dead, so too does counterfactual mourning trouble the hegemonic logic of extractivism by positing other ways of being in relation to human and nonhuman life, embodied in the person who fought to make it a reality: "Samir vive."

In what follows, I analyze the aesthetics of counterfactual mourning in response to the murder of land defenders in Mexico. I frame my study with a discussion of Alfredo Joskowicz's film *El cambio* (The change, 1975), a pioneering depiction of state-sponsored violence against environmentalists. *El cambio* concludes with murder, illustrating how murder is traditionally emplotted as an end. By contrast, counterfactual practices of mourning contest this emplotment, scrambling murder's conclusivity. I look at how visual forms of counterfactual mourning like graffiti and murals invite passersby to enter into new kinds of imaginative relation with territory as well as with the practice of land defense. I then turn to Naomi Rincón Gallardo's performance cycle *Una trilogía de cuevas* (A cave trilogy), to examine how it extends this invitation in a way that foregrounds the queer and unruly embodied pleasures of land defense.

BARE VIOLENCE AS A MODE OF PRODUCTION IN *EL CAMBIO*

Mexico's first explicitly environmentalist feature film, Alfredo Joskowicz's *El cambio* (1975), culminates in murder.[10] Filmed in 1971, this antecedent of contemporary representations of violence against land defenders puts forth an argument that processes of industrialization are made possible by state-sponsored violence. The film follows two countercultural artists as they seek refuge from Mexico City's smog in what they initially perceive to be untouched nature, traveling to Tecolutla, Veracruz, to rough it on the beach. The fantasy of escaping modernity's grime is quickly upended when the protagonists discover that the beach is being polluted by a factory, threatening the local fishing economy. After unsuccessfully attempting to organize the community's impassive fishermen, the two *chilango* hippies take the fight upon themselves. They interrupt the factory's inauguration and throw sludge on a company executive after he lauds the factory's "desarrollo favorable" (favorable development) of the town. Drenched in oil and humiliated by the laughing crowd, the executive orders a group of uniformed armed men to pursue the duo to the beach, where they execute the environmentalist hippies in the denouement. An understudied film, *El cambio* unwittingly stages first-wave environmentalism's problematic imposition on "ignorant," "disinterested," or "oblivious" rural communities by educated urban outsiders. What is of most interest to me here, however, is not to critique the film's replication of counterculture's raced, gendered, and classed flaws, which have been comprehensively outlined by Rebecca Janzen.[11] Rather, I want to underscore that its culminating scene explicitly dramatizes how bare violence is used to foreclose dissent to the extractivist paradigm.

El cambio's bloody denouement has been interpreted as an allusion to the 1968 Tlatelolco massacre, which at the time could not have been directly depicted on screen without censorship.[12] The confrontation between the idealistic young protagonists and the uniformed police draws an unambiguous parallel to this event. After their activist stunt, the two youthful hippies exuberantly abscond to the beach, collapsing into a pile behind a mound of coconuts, laughing and out of breath, aware they might be followed, but not expecting to be met with violence. The uniformed police arrive in pursuit, and in a high angle shot, four armed men loom above the protestors' makeshift refuge. After the protagonists reluctantly emerge, each man is successively shot in the stomach. The scene is intimately staged. The men are executed at close range; victim and victimizer sustain eye contact. In response to the men's unbroken gazes, the head of the security detail repeatedly barks, "¿Qué me ves, hijo de la chingada?" (What are you looking at / What do you see in me, son of a bitch?), receiving no reply. This lingering reiterative question implies that Tlatelolco marked a definitive shift, or titular "cambio," in how the state and the PRI were viewed by the middle class. By deterritorializing the 1968 massacre from the urban to the rural, *El cambio* situates the political stakes of the 1968 student movement as not containable to a specific time or place.

Yet *El cambio* is not just allegorical pretense, a way for Joskowicz to treat Tlatelolco on screen. The concluding execution of the young environmentalists can also be read literally, as the founding cinematic articulation of state-sponsored terror against land defenders. As the camera pans away from the protagonists' corpses and out to the polluted ocean, *El cambio* illustrates Achille Mbembe's observation that terror is a mode of production essential to "processes of appropriating economic resources."[13]

In narrative terms, death's decisiveness means that it is often equated with closure. We see this at work in *El cambio*, which maps execution onto denouement. The film's concluding scene provides a definitive end to the conflict in Tecolutla; the men's killing truncates opposition to developmentalism and sustains the status quo. This emplotment suggests that state violence forecloses political dissent and that it quashed first-wave environmentalism in Mexico before it could gain traction. In the filmic universe of *El cambio*, death is a resolution that relieves suspense, meeting the audience's desire for a clear outcome, if not a happy ending. Yet closing the film on the scene of execution suggests that the conflict in Tecolutla is over, wresting attention away from the aftermath: the polluted beach and the fishermen who must contend with the effects of industrial contamination. Even when land defenders are murdered, conflicts continue, whether through ongoing protest, legal action, or the memory of the displaced. Therefore attending solely to death can overshadow the processual nature of antiextractivist resistance. In the analysis that follows, I demonstrate how counterfactual aesthetic and rhetorical strategies avoid this pitfall by reframing the murder of land defenders, not as the end of the story, but as a point of departure for continued imaginative work and political struggle.

Fast-forwarding to today, of the countless films and novels that depict violence in twenty-first-century Mexico, very few explicitly explore its connection to land defense and extractivism. This is starting to change, as films like *Huachicolero* (The gasoline thieves, 2019, directed by Edgar Nito) and *Noche de fuego* (Prayers for the stolen, 2021, directed by Tatiana Huezo) and novels like Fernanda Melchor's *Temporada de huracanes* (Hurricane season, 2017) narrate extractivist ecosystems of violence. Yet the overall elision of extractivism in contemporary cultural treatments of violence in Mexico replicates an omission identified by Oswaldo Zavala in mainstream discussions of the so-called war on drugs. Drawing on the conclusions of journalists Ignacio Alvarado, Dawn Paley, and Federico Mastrogiovanni, Zavala observes that "'the war on drugs' is the public name for political strategies that displace entire communities and appropriate and exploit natural resources that otherwise would be out of reach for national and transnational capital."[14] The drug war justifies the use of force against populations throughout Mexico inhabiting territories ripe for concessions, and increased violence runs parallel, not to an escalation in cartel activity, but to neoliberal reforms that have facilitated the privatization of biodiverse territories through militarization.[15] As in *El cambio*, state forces secure the interests of capital over those of citizens. This has led Cristina Rivera Garza to describe contemporary Mexico as a "visceraless state," a state that establishes "disembowled relationships" with its constituency.[16] This is a state that values "the supremacy of profit above life" and rescinds "its responsibility for the care of its constituent's bodies."[17] This indifferent and neglectful state allows and perpetuates horrific violence, literal and figurative, so long as its economic imperatives are met.

LAND DEFENSE AS ENVIRONMENTALISM AND *COMUNALIDAD*

Across Latin America, violence against land defenders has escalated along with the expansion of the extractivist frontier.[18] Maristella Svampa has explained that at the turn of the twenty-first century, advanced capitalism's insatiable demand for raw materials and energy led to a boom in global commodity prices that encouraged administrations across the political spectrum to intensify export-oriented extractive production. This economic reprimarization incentivized the extension of the commodity frontier into "new territories [that] were previously considered unproductive or not valued by capital" and now could be exploited via new technologies.[19] In Mexico, a slew of neoliberal reforms to energy and mining laws in the wake of the oil crash in the 1980s opened these industries up to transnational participation. Illustrative of this codification of extractivist priorities (or what Svampa calls the "commodity consensus") are the 1992 reforms to Mexico's mining law under PRI President Carlos Salinas de Gortari, which declared that "the exploration, exploitation, and processing of minerals . . . will take precedence over any other use . . . of the land."[20] What followed were decades of government concessions granted to transnational companies on communal and Indigenous lands without informed consent.

As a result, conflicts surged between affected communities and extractivist industries, with the latter buttressed by state protection and impunity. The people who resist the imposition of these projects, like Samir Flores, pose an obstacle to capital. The threat they represent is ideological, since land defenders offer a competing understanding of territory oriented around participatory becoming rather than possession.[21] Mexico has one of the highest global rates of violence directed at land defenders, broadly understood as anyone involved in efforts to protect a specific territory from its commodification. The murder of land defenders represents only "the tip of the iceberg," given that "for every defender murdered, thousands more face direct violence, threats and psychological intimidation, and more invisible cultural and structural violence."[22] According to the Centro Mexicano de Derecho Ambiental (CEMDA, Mexican Center for Environmental Law), between 2012, when they began tracking cases, and 2022, environmental defenders reported more than 850 attacks.[23] An accurate accounting of the *cifra negra*, or hidden figure of unreported crime, would likely yield higher numbers, given that impunity and intimidation disincentivize reporting: 94 percent of crimes against land defenders go unreported in Mexico.[24] Even without clear statistics, it is evident that aggression against land defenders in Mexico is on the rise. Fifty-four land defenders were murdered in 2021, and thirty-one in 2022; a third of all victims were forced disappearances, and forty percent were Indigenous. In 2022, CEMDA documented 197 acts of aggression against land defenders, an 82 percent increase from 2021. Forty-five percent of these attacks, CEMDA found, were perpetuated by the Mexican state.[25] Land defenders were murdered every month of the year in 2021 and 2022, making Mexico one of the most dangerous countries for land defenders in the world.[26]

The scale of violence inflicted against land defenders boggles the mind, its seemingly infinite repetition a "pedagogy of cruelty" that Rita Segato explains produces the intended effect of social atomization and desensitization, leading "to the low levels of empathy that are indispensable for predatory enterprises."[27] I offer one case by way of an example. When I began working on this chapter in January 2021, Chatino leader Fidel Heras Cruz was assassinated in retribution for his leadership of Consejo de Pueblos Unidos por la Defensa del Río Verde (COPUDEVER, Council of United Peoples for the Defense of Río Verde), a group resisting planned sand mining and hydroelectric projects on Río Verde, a body of water that sustains forty-three Chatino, Mixe, mestizo, and Afro-descendent communities in the state of Oaxaca. By March, as I drafted these lines, four other members of COPUDEVER had been murdered: Raymundo Robles Riaño, Noel Robles Cruz, Gerardo Mendoza, and Jaime Jiménez Ruiz. Outside of Oaxaca, the murders received scant attention.[28] In the context of widespread violence, the murders of land defenders too often get lost in the shuffle.

These invisibilized murders are akin to slow violence in their accretive, paralyzing effects. Rob Nixon has compellingly theorized "slow violence" as a framework that expands how we think about violence, pointing out that many forms of environmental violence are attritional or accretive, rather than instantaneous.[29] But while Nixon identifies the representational challenge of temporally latent violence, Cristina Rivera Garza identifies its inverse: the traumatizing effect of omnipresent violence. Twenty-first-century Mexico, Rivera Garza writes, is living through "one of the most chilling spectacles of contemporary horror." Disappearance has become ordinary and widespread, manifest through feminicide and the killings of migrants, journalists, and human rights advocates. The effect of witnessing this horror on a daily basis, whether on the news or in one's community, Rivera Garza explains, is paralyzing. "Bewildered and immobile, the horrified are stripped of their agency.... They stare, and even though they stare fixedly, or perhaps precisely because they stare fixedly, they cannot do anything."[30] The task of mourning, then, is to produce ways of agential seeing. This task echoes *El cambio*'s reiterative query "¿Qué me ves?," with its dual meaning of "What are you looking at?" and "What do you see in me?," an interrogation that identifies both the gaze's unsettling effect on power and its interpretive potential to reorient sight.

Building off Zavala and Paley's argument that violence in twenty-first-century Mexico has been frequently misunderstood as originating in the drug war rather than in the struggle over natural resources—a misdirection that benefits the commodity consensus—the murder of land defenders on the front lines of the extractivist frontier demonstrates the extent to which issues of land are inextricable from questions of race, gender, class, and survival. While horrific events like Ayotzinapa and the #NiUnaMenos campaign against feminicide have reached great levels of public visibility due to the concerted efforts of activists and artists, the murder of land defenders, many of whom are Indigenous, has yet to reach this level of recognition. It bears asking why not. The ethnic and rural makeup of the victims of land defense may account for this invisibility, as Indigenous peoples have long been framed by the state and the media as obstacles to national development, and conflicts surrounding land defense are often misrepresented as localized disputes. These racist misinformation strategies divert attention from the extractivist entities that perpetuate violence to quash opposition and muddy the waters by attempting to paint both sides of land disputes as equally at fault. The mainstream invisibility of extractivist violence is not therefore the result of public disinterest, but the desired outcome of misinformation campaigns waged by public and private forces aligned with extractivist business.

While not all land defenders are Indigenous, it is imperative to underscore the role that ethnicity plays in these conflicts, as it has in others. Even in the highly visible case of the Ayotzinapa massacre, anthropologist Mariana Mora explains, journalists primarily identified the victims as campesinos and only infrequently noted that several victims were Naa Savi, Me'phaa, Nahua, and Huave.[31] This erasure of ethnic difference by the media mirrored the federal response, which marginalized victims' families by not offering them interpreters during meetings about the case. Mora argues that it is important to foreground racism in scholarly and public conversations about forced disappearance, because it opens up new demands that can be made to the state and new ways of thinking about the law, justice, and mourning as they are expressed from divergent cultural positions.[32]

Along these lines, Ayutla Mixe linguist, activist, and writer Yásnaya Elena Aguilar Gil has succinctly explained: "nuestro ambientalismo se llama defensa de territorio" (our environmentalism is called land defense).[33] In this concise phrase, first issued as a tweet and subsequently developed into an op-ed for *El país*, Aguilar Gil identified the schism between environmentalism emerging from urban, bourgeois sectors, which has been frequently co-opted by the interests of capital, and Indigenous efforts to defend the land as part of a strategy of cultural survival. (The former is neatly embodied by the two protagonists in *El cambio*, who impose their ideas of environmental purity on the rural fishing community.) The schism between these two veins of environmentalist thought is long in the making, informed by the racialized and racist marginalization of Indigenous politics as anathema to urban environmental pursuits. This separation has undermined coalitions between the two groups, to the benefit of the status quo.[34]

Beneath this surface schism is the reality that, according to Víctor Toledo, David Garrido, and Narciso Barrera-Bassols, "the defense of territory and territoriality is the most visible programmatic feature of the varied environmental struggles and movements of Mexico and Latin America."[35] *Territory* is a broad term with multivalent meanings, but generally indexes the material space through which social relations are produced; these social relations, in turn, actively produce territory as a commons.[36] In contrast with capitalist understandings of land as a passive space related to through possession, territory is understood by Amerindian philosophies as a woven fabric composed by human and nonhuman participants.[37] Mixe philosopher Floriberto Díaz explains that while in Western thought community is additive (the sum of individuals), for Indigenous peoples, community is a geometric relationship assembled around the land.[38] Territory formulates the ontological basis of self and *comunalidad*, as evident in the episteme explicated by Díaz: "1. Donde me siento y me paro > 2. En la porción de la Tierra que ocupa la comunidad a la que pertenezco para poder ser yo > 3. La Tierra, como de todos los seres vivos." (1. Where I sit and stand > 2. On the part of the Earth occupied by the community that I belong to in order to be myself > 3. The Earth, like all living beings.)[39] This positional understanding of the self always in relation to others through territory forms the basis of Mixe political thought, as well as of many rural communities in Mexico—even those who do not understand themselves to be Indigenous, but whose cultural beliefs, as Guillermo Bonfil Batalla famously posited in *México profundo*, are inflected by Mesoamerican concepts.[40] Land defense, understood this way, both is coterminous with environmentalism and also exceeds it, gesturing to a more expansive account of relationality between space and people formulated around geometric belonging.

Given the centrality of territory to *comunalidad*, critics have noted the abyssal disconnect between state support for Indigenous cultural production and the escalation in concessions that jeopardize those same peoples. Yucatec Maya poet Pedro Uc Be sums up this situation: "Las instituciones de gobierno que celebran la lengua maya en el marco del día de las lenguas maternas, tendrán legitimidad, si y sólo si se pronuncian por la cancelación de los megaproyectos que invaden nuestro territorio maya en donde creamos nuestra lengua y cultura maya. ¡No queremos consulta, queremos territorio!" (State institutions that celebrate the Maya language in the context of International Mother Tongue Day will only obtain legitimacy if they also advocate for the cancellation of megaprojects that invade Maya territory where we create our language and culture. We don't want consultation, we want territory!)[41] The state's superficial support for Indigenous culture at the expense of Indigenous politics signals how cultural extractivism unfolds in tandem with material extractivism, as both Indigenous culture and territory are mobilized as commodities.[42]

Of course, it is important not to homogenize Indigenous politics when it comes to developmental projects. Some communities are adamantly opposed to megaprojects, while others adopt a posture of negotiation. Ana Matías Rendón explains, "Los pueblos nunca han sido homogéneos. No existe, no ha existido, y tal vez, jamás exista una unidad 'indígena' [. . .] Cada pueblo es un sistema epistemológico [. . .] Por ello, cada pueblo, tiene sus razones para aceptar o rechazar un proyecto." (Indigenous peoples have never been homogenous. There is no "Indigenous" unity, it has never existed.

... Each people is an epistemological system. ... Because of that, each people has their reasons to accept or reject a project.)[43]

Nonetheless, violence against land defenders underscores the threat that Indigenous approaches to territory pose to capital. Zoque writer Mikeas Sánchez's poem "Jujchere' / ¿Cuánto vale?" (in translation by Wendy Call as "What Is It Worth?"), in her sardonically titled book *Jujtzye tä wäpä tzamapänh'ajä / Cómo ser un buen salvaje* (How to Be a Good Savage, 2019), illustrates this point. Inverting the classic association of indigeneity with barbarism, her poem opens with a series of propositions made by capital:

>Te' yajkuyis'nhkyowinastam, tä' näjmatyamba:
>Mij' nhkajkabyatzi sone'ruminh'jinh
>te' tzujtzibä' mij' dzajp,
>mij' dzäjkpujtabyatzi saxapyä' maa'räjk
>uka' dyaj täjkäbya mij' nhkotzojk'omoram.
>Tumä'millon tzujtzirambä'ruminh
>wäkä' jambää' jujche kasäyajpa mij' uneram
>poyapajk oñdyujomo.

>Los amos de la barbarie nos dicen:
>Te ofrezco una cuenta millonaria
>a cambio de tu cielo azul,
>te construyo un hermoso supermercado
>a cambio de tus montañas.
>Un millón de dólares
>por la sonrisa de tus hijos
>que corren bajo la lluvia.

>Those masters of barbarity tell us:
>I'll give you a millionaire's bank account,
>in exchange for your blue sky,
>I'll build you a nice supermarket
>in exchange for your mountains.
>One million dollars
>for your children's smiles
>as they run in the rain.[44]

After revealing the absurdity of extractivism's exchange rate, the poetic voice makes no counter, instead responding with dismissive laughter and questions of their own. Note the contrast between the singular first person associated with the "masters of barbarity" and the collective "we" of the Mokayas (the people of the corn), the epistemic difference between the two demarcated by divergent verbs, "telling" versus "asking":

> Mokayas'tam mij' nhkosijktatymbatzi' mij' dzame',
> mochirambä'uneis myuxajpabände,
> jujche te' tuminh yatzyäyubä wakas'tinhajpa,
> dä' nhkätpak te' Tzuan'.
> Mokayas'tam mij' nhkämetztambatzi' mijtam',
> yajkuyis' nhkyowina'ram.
> ¿mij' banku'omorambä' tuminh'jinh
> mujspa'a yajk' wyrujatyamä
> Tzusnäbajkis'xasa'ajkuy?

> Los Mokayas nos reímos de su ignorancia,
> hasta los niños más pequeños
> saben que la fortuna se convierte en boñiga
> pasando la línea del Tzuan.
> Los Mokayas les preguntamos a ustedes,
> amos de la decadencia.
> ¿Una cuenta millonaria
> será suficiente para devolverle
> la alegría a nuestros muertos?

> We Mokayas laugh at their ignorance,
> even the smallest children
> know that money turns to manure
> when you pass over to Tzuan.
> We Mokayas ask you,
> the masters of decay.
> Is a millionaire's bank account
> enough to bring back
> the laughter of our dead? [45]

The afterlife, or Tzuan, a realm in which money is rendered manure, destabilizes extractivism's recompensatory model. In death, Sánchez's poem implies, the logic of capitalism is undone, because in death money has no value. It is perhaps for this reason—the afterlife as the space where capitalism is unmoored—that the murder of land defenders becomes a flashpoint in establishing the stakes of counterfactual mourning.

TAKING UP SPACE: COUNTERFACTUAL MOURNING IN STREET ART

In the late summer of 2021, I flew to Oaxaca to complete some research, speak with artists, and visit La Jícara, an independent bookstore. As I hurried to deplane, I noticed a huge banner hanging at the entrance to baggage claim welcoming visitors to Oaxaca, an advertisement for Compañía Minera Cuzcatlán, a subsidiary of the Canadian Fortuna Silver Mines. The same ad campaign was prominently visible at the roundabout where taxis and cars exit the airport grounds, the first message that every air-bound visitor to Oaxaca sees upon leaving the airport. Minera Cuzcatlán has long employed this sort of Oaxaca-as-mine branding strategy to target tourists and locals alike. In 2019, banners at the airport featuring smiling miners read, "Somos industria oaxaqueña" (We are Oaxacan industry) just a month before a fatal accident suspended operations at one of the company's mines.[46] Minera Cuzcatlán has also been a regular sponsor of the annual Guelaguetza Festival, Mexico's largest Indigenous cultural celebration, and in 2018, it sponsored Oaxaca FilmFest, the city's film festival.[47] Through these prominent corporate sponsorships, Minera Cuzcatlán self-fashions itself as a force aligned with Indigenous peoples and indispensable to the infrastructure of Oaxacan culture.

In reality, since 2006, Minera Cuzcatlán has been in prolonged conflict with at least one hundred Zapotec families from San José del Progreso who oppose its operations.[48] It forcibly evicted protestors blockading the San José mine in 2009 with the help of federal and state police. In 2012, Bernardo Vásquez Sánchez, the leader of the activist opposition group Coordinadora de Pueblos Unidos del Valle de Ocotlán (COPUVO, United Peoples' Network of Ocotlán Valley), was assassinated in his car. With him was Rosalinda Dionicio, a lawyer and activist (and the protagonist of *Resiliencia tlacuache* by performance artist Naomi Rincón Gallardo, analyzed further ahead), who was gravely wounded. Then, in 2018, heavy rainfall caused the San José mine's open-air waste pool to overflow, spilling radioactive tailings into a nearby river and contaminating local water. Within this context, Minera Cuzcatlán's public positioning of itself as inextricable from Oaxaca's biggest cultural events is a defensive response to a decade of organized resistance and a move that signals the centrality of public space as an arena in which extractivism is normalized or called into question.

In the face of terror buttressed by sophisticated PR campaigns, public mourning becomes a way to produce a world defined by relation, rather than outside of it. "Deep sorrow," Rivera Garza writes, "binds us within emotional communities willing and able to face life anew, even if it means,

FIGURE 2.1. "Samir vive" (Samir lives), Mexico City, 2021. Photograph by the author.

or especially when it means, radically revising and altering the world we share."⁴⁹ This revisionary production of social space adopts different forms, including localized blockades, ephemeral street protests, and legal action. These interventions disassemble extractivism's performative identification with place—exemplified by Minera Cuzcatlán's branding of Oaxaca-as-mine—producing instead emotional communities around territory and the dead.

That summer, walking around Mexico City, I noticed "Samir vive" graffitied and painted across the sides of buildings, honoring the aforementioned Nahua activist and leading opponent of Proyecto Integral Morelos (fig. 2.1). Likewise, in Oaxaca, I came across the stenciled face of Tomás Martínez Pinacho, an antimining activist and leader of Frente Popular Revolucionario, assassinated in 2020 (fig. 2.2). These spectral traces of the dead inscribed in the realm of the living reinstate them as public figures. The face interpellates passersby, evoking what Emmanuel Levinas has described as an invitation to relation.⁵⁰ As an apparatus of political agency, visual inscriptions reaffirm the right to public space on behalf of those who have been forcibly removed from it as well as on behalf of the community and cause

FIGURE 2.2. "Crimen de estado," anonymous stencil of Tomás Martínez, Oaxaca City, 2021. Photograph by the author.

that survive them. This double move renders the victim simultaneously singular and universal: a person with an individual experience who also personifies a cause and is metonymic of a specific territory.

Whether reproduced on city walls, as posters in street demonstrations, or for social media campaigns, representations of murdered land defenders crystallize around a delimited set of characteristics: face, name, and slogan. These pared-down aesthetics of remembrance echo strategies that were first popularized during the 1960s and '70s in the Southern Cone in response to the forcibly disappeared desaparecidos and have since been mobilized throughout Latin America to contest impunity. An emblematic recent example is the activist response to the 2014 massacre of the forty-three *normalistas* in Ayotzinapa. The reiterative diffusion of the forty-three student victims' names and faces in countless public spaces gave form to an omnipresent claim against the state. Pasted on walls around the country, the students' portraits functioned, Laura González-Flores explains, as "effective proxies ... symbolically [returning] them to shared public space."[51] This visual presence helped mobilize the living, who maintained public pressure on authorities eager to sweep the case under the rug.

Applied to land defender victims, the distilled visual representation simplifies highly complex localized struggles into an immediately recognizable idiom of impunity, connecting these cases with other ongoing social justice movements, like Ayotzinapa and #NiUnaMenos. Whereas Ayotzinapa captured public attention because of its spectacular singularity, like the victims of feminicide, murdered land defenders are cumulatively great in number, but their deaths have a hard time gaining national visibility because of their atomized occurrences in far-flung contexts and because of misinformation campaigns that criminalize victims or misattribute their death to localized disputes or the drug war. In this sense, the economical and even formulaic presentation of victims aims for familiarity in a move that brings to mind Susan Sontag's description of how political posters borrow the lesson of simplicity from advertising.[52] Even when victims are unfamiliar to the passerby, succinct phrases like "Samir vive" trigger immediate recognition: in this case, identifying the dead through the counterfactual assertion of their life. Through reiterability, the slogan places the individual victim within a network of similar cases—"Berta vive," "Tomás vive," "Bety vive"—a discursive assemblage that figures terror as both intimate and seemingly infinitely repeatable.

The purpose of counterfactual slogans like "Samir vive" and the distilled representation of land defenders in public space is not pedagogical—to teach the public about the victim or their specific cause. In fact, specifics are notably absent. Rather, the goal is affective: to spark a form of mourning that actively negates terror's intended eradication of space. The first name address fosters an easy familiarity that is reinforced spatially through quotidian placement, so that one encounters the names of the dead on one's daily route through the city.

At the same time, the phrase "Samir vive" abstracts beyond the individual through the encoded allusion to the Zapatista mantra "¡Zapata vive, vive, la lucha sigue, sigue!" (Zapata lives, lives, the struggle continues, continues!), which itself drew on the looming historical figure of Emiliano Zapata. Establishing a direct lineage between contemporary land defense struggles and the Mexican Revolution's demands for land and liberty, "Samir vive" appeals to leftist political ideals and the fight for Indigenous autonomy at the same time that it makes a more mainstream, nationalist appeal through the figure of the founding father. A similar move is made through form in the stenciled visage of Martínez (fig. 2.2). The stencil emulates the artisanal appearance of woodblock prints, a style commonly used in political posters in the wake of the Mexican Revolution. Because extractivism has been so pervasively associated with national development and jobs, and land

FIGURE 2.3. "Justicia: Samir es semilla" (Justice: Samir is a seed), Mexico City, 2021. Photograph by the author.

defenders so frequently delegitimized through tactics like criminalization, their aesthetic and discursive association with revolutionary ideals deploys nostalgic nationalism to build expansive political coalitions among urban, mestizo, and transnational publics. This invitation to solidarity is made by triggering an assemblage of ideals that bloom in the mind of the passerby: land, liberty, impunity, peasantry, indigeneity, maize, autonomy (fig. 2.3). The downside of this associative move is that it reorients social memory of land defenders away from the specific struggles they embodied.

Street art and social media campaigns that represent murdered land defenders tend to treat the victim with reverence, depicting them as a martyr or heroic figure. In large part this is an effort to counteract negative mediatic representations that discursively conflate, as Mikeas Sánchez explains, "el indígena, el pobre o el que protesta [. . . con] un vándalo" (the Indigenous person, the poor person, or the protestor . . . [with] the vandal).[53] Painting protestors as the problem, the media tends to focus on blockades or the destruction of private property rather than the concrete complaints behind these actions. Yet the drawback to responding to this delegitimization through hagiographic tactics—such as the halo that murdered land defenders are often endowed with, which renders them angelic martyrs—is that it recurs to the model of the "legitimate" or "ideal" victim. In her study of the "legitimate" victim, Sandra Walklate argues that this reductive model, albeit effective in cultivating empathy, allows some to be seen as "deserving victims" due to their conduct, while others are "viewed as undeserving victims who may never be labeled as victims" at all—for example, those who might negotiate with capital or labor in the extractive industry but still suffer its effects.[54]

Despite these caveats, ultimately the inscription of the images and names of the dead powerfully restore them to social space. In *The Production of Space*, Henri Lefebvre explains that space is socially produced: "each living body is space and has its space: it produces itself in space and it also produces that space."[55] And what of the dead? Drawing on Lefebvre, we can see the murder of land defenders as an eradication of space that is necessary to produce the space of extractivism—Oaxaca-as-mine—without room for dissent. The body's spectral, counterfactual restitution via visual representations in public space subverts this spatial consolidation. It counteracts the intended effect of murder by reinstating the victim's material appearance in the social sphere, reminding the public of their political presence and inviting forms of social relation beyond extractivism.

These observations hold true for the images created in the wake of the murder of Alberta "Bety" Cariño, a Mixtec human rights defender and the director of the Centro de Apoyo Comunitario Trabajando Unidos (CACTUS, Center of Communal Support Working Together). Cariño was killed in 2010 by a government-backed paramilitary group, along with Finnish international observer Jyri Antero Jaakkola, when their humanitarian caravan came under fire as they attempted to access the autonomous Triqui municipality of San Juan Copala, which had been deprived of basic provisions by an ongoing paramilitary blockade. Due to the intersectional nature of her activism and the brazenness of her murder, Cariño, whom I will discuss again shortly for her role in Rincón Gallardo's performance piece, became in Oaxaca an icon for Indigenous land and feminist struggles, as well as the struggle against impunity.

Like in the aforementioned examples, activist imagery honoring Cariño has focused on her face, usually set in an expression of dignity or happiness against a backdrop of maize. The image's emotional appeal is amplified by accompanying phrases like "Bety vive, la lucha sigue" (Bety lives, the fight goes on) and "Nuestras ancestras, semillas de nuestra lucha" (Our ancestors, seeds of our fight), which make a temporal claim that Cariño's death did not mean the end of her struggle.[56] The counterfactual expression "Bety vive" rejects the imposition of extractivism as the only possible world, instead asserting the subjunctive existence of other worlds: worlds in which

FIGURE 2.4. Lapiztola, *Sembremos sueños y cosechemos esperanzas*, 2015. Courtesy of the artists.

Cariño lives on through the actions of others and worlds yet to come that germinate from her past.

This claim to futurity is underscored in a 2015 mural honoring Cariño by the public art collective Lapiztola (Roberto Vega and Rosario Martínez), their name a portmanteau of *lápiz* (pencil) and *pistola* (pistol), painted on the side of Museo Belber Jiménez in the historical heart of Oaxaca City (fig. 2.4). Instead of Cariño's face, the mural *Sembremos sueños y cosechemos esperanzas* (Let Us Plant Dreams and Harvest Hope) features a young Indigenous girl wearing an embroidered huipil, rendered in black and white. Seated atop a bed of flowers, the supersized stenciled figure holds a blood-red heart in her hands and looks meaningfully into the distance. A scroll that extends across her body contains a quote by Cariño: "Hermanos, hermanas, abramos el corazón como una flor que espera el rayo del sol por las mañanas, sembremos sueños y cosechemos esperanzas, recordando que esa construcción solo se puede hacer abajo, a la izquierda y del lado del corazón" (Brothers, Sisters, let us open our hearts like a flower that waits for the morning sun, let us plant dreams and harvest hopes, remembering that this project can only be undertaken beneath, to the left, and alongside the heart). This homage makes an emotional plea to passersby in Oaxaca's touristy downtown, trafficked by a cross-section of wealthy foreigners, domestic tourists, backpackers, and urbanites; it appeals to their sense of obligation to "open their hearts" and call for meaningful change in a region that is marketed for

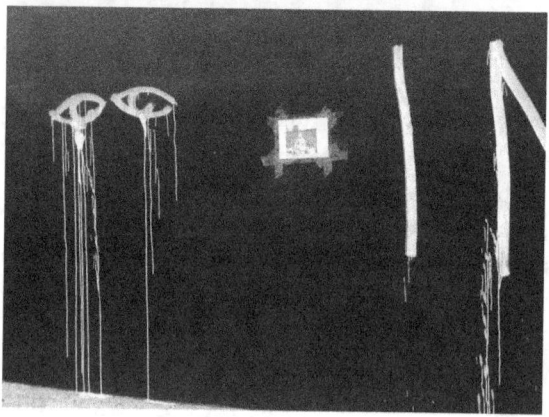

FIGURES 2.5, 2.6, and 2.7. Lapiztola's *Sembremos sueños* mural was painted over by municipal officials in 2015. Subsequently, an anonymous intervention appeared on the site. It included a set of eyes and the graffitied phrase "I miss you" in English, and a miniature print of the original mural in a makeshift frame. Courtesy of Lapiztola.

its cultural diversity and yet whose Indigenous communities are the targets of extractive megaprojects and political oppression.

To outsiders visiting Oaxaca, a city renowned for its vibrant public art scene, the mural may not be immediately legible as a memorial to Cariño. Instead of depicting Cariño's face, as is often the case in public homages to land defenders, the artwork encodes its homage within a seemingly apolitical image of a young Indigenous girl. The malleability of the anonymous representative figure allows the image to be read and consumed by different publics on different registers. For locals it points toward ongoing political projects, like the fight for Triqui autonomy and the violent repression of these dreams; it invites a second look that registers Cariño as the quote's author who endows the girl's image with new meaning: a symbol of the future of the Triqui project for autonomy that Cariño died defending. Yet for foreign tourists, the *Sembremos sueños* mural may carry none of these political associations and instead may be consumed as just another instance of Oaxaca's vibrant street art scene, ready-made for selfies.

While the aesthetics of *Sembremos sueños* map in many ways onto a recognizable visual idiom of Oaxacan-ness, the encoded reference to Cariño— and thus to impunity and dispossession—made it nonetheless politically charged. In late 2015, the municipality painted over Lapiztola's *Sembremos sueños* mural on the pretext that it flouted ordinances regulating the historical center's color scheme and "identity" (fig. 2.5).[57] Although public art is inherently ephemeral, the removal of a vibrant mural alluding to Cariño's assassination for her support of Triqui autonomy exemplifies how Oaxaca's tourist branding as a safe and convivial city known for its well-preserved, accessible Indigenous cultures relies on the censorship of these realities, including their eradication from public space.

Yet, as with many cases of state censorship, the mural's effacement merely garnered it more attention. In the wake of its erasure, the site became a makeshift altar, "a site of remembrance of the image," as Marisol Rojas observed about the incident.[58] Behind the thin coat of blue paint the municipality used to cover the stencil, the girl's face and body were still faintly visible, a shadow that drew attention to the mural's absence. An anonymous graffiti artist tagged the repainted wall with a pair of eyes and the words "I miss you" in English in white paint. Someone else affixed a miniature print replica of the *Sembremos sueños* mural to the wall and placed candles underneath (figs. 2.6 and 2.7).

The resignification of *Sembremos sueños* after its censorship into a makeshift altar for Cariño as well as for the continued silencing of her murder by the state illustrates how public practices of mourning nourish imaginative political association with the body-territory even within a context of censorship and impunity. In contrast with the PR campaigns by extractivist entities like Minera Cuzcatlán, which obtain hypervisibilty by associating themselves with large cultural events, unauthorized interventions that insert reminders of murdered land defenders into public space recreate that space as a participatory site of imaginative engagement with antiextractivist political projects. Alice Driver deems this sort of "witnessing through testimony . . . constructed on a specific or meaningful geography" a form of "ecotestimonio."[59]

In the decade that has passed since Cariño's murder, her family, community, and network of political allies have circulated Cariño's face and name with great diligence to render her familiar to the public. These forms of remembrance still take up public space today, for instance, through signage in Oaxaca's zócalo that decries the continued displacement of the autonomous Triqui municipality of San Juan Copala, where Cariño and the humanitarian caravan were headed when they came under fire. Cariño's murder, the perpetrators of which have not been brought to justice for ten years, is inextricably linked to broader processes of Indigenous dispossession. As the Red Mexicana de Afectadas/os por la Minería (REMA, Mexican Network of People Affected by Mining) put it in their annual call memorializing Cariño's death: "Sabías claramente que la defensa del territorio es el camino, que no hay más a donde ir. [. . .] Tu asesinato no solo evidencia la cobardía, sino los contubernios, las alianzas de grupos de poder, la podredumbre de quienes están protegiendo el saqueo y la mercantilización de los bienes comunes" (You knew clearly that the defense of the land is the way, that there is nowhere else to go. . . . Your assassination evidences not only cowardice, but the alliances of power and conspiracies among those who protect the pillaging and commodification of the commons).[60] Public art plays an essential role in counterfactual mourning: it stands in for the murdered land defender, multiplying their presence across a variety of spaces, visually materializing their continuation—"Bety vive"—and rhetorically inviting passersby to carry their political project into the future.

In addition to the aforementioned murals and posters, Cariño's memory has been kept alive as a seed through a variety of projects, including the publication of her poetry by independent press El Rebozo, songs ("El corrido de Bety Cariño" by Félix García), a graphic novel (*Sembrando sueños, cosechando esperanzas* by Spanish illustrator Ruma Barbero for the Basque NGO Mugarik Gabe), and multiple documentaries (including the CACTUS production *Bety Cariño, tejedora de esperanza*, Iván Castaneira's *Homenaje a Bety Cariño*, and *Murha Meksikossa / A Mexican Murder Story / Muerte en México* by Finnish documentarian Simo Sipola, dedicated to Jyri Jaakkola).[61]

Cariño's memory also inspired Naomi Rincón Gallardo's performance piece *El viaje de formol*, which I discuss next.

IRREVERANCE IN NAOMI RINCÓN GALLARDO'S *UNA TRILOGÍA DE CUEVAS*

Like urban art interventions that create space for the dead by interrupting the passerby's visual field and inviting engagement, performance art also creates space for the body-territory through enactment or "doings" of diverse bodies brought together around shared space. This is the case in performance artist Naomi Rincón Gallardo's *Una trilogía de cuevas*, a three-part multimedia series about extractivism and dispossession in Mexico. The trilogy includes *El viaje de formol* (*The Formaldehyde Trip*, 2017), *Sangre pesada* (Heavy blood, 2018), and *Resiliencia tlacuache* (Opossum resilience, 2019). Originally commissioned by the San Francisco Museum of Modern Art as part of its Performance in Progress series (cocurated by Frank Smigiel of SFMOMA and Ani Rivera of Galería de la Raza), each chapter of *Una trilogía de cuevas* has been screened in museums or performed live, bringing counterfactual mourning to new publics.[62] Collectively, the trilogy highlights Indigenous women's leadership in land defense and materializes queer horizons unbound by patriarchy and accumulation. These feminist "counterworlds," as Rincón Gallardo describes them, are transtemporal, nourished at once by Mesoamerican myth, Indigenous activism, and queer theory in order to engender "worlds opposite to the present for a future that is not yet here."[63] In this sense, Rincón Gallardo's speculative trilogy is another example of subjunctive aesthetics. It complements and extends the counterfactual negation of death's finality in activist phrases such as "Samir vive" and "Quien lucha por la vida nunca muere" (Those who fight for life never die) by enacting what the parallel worlds cultivated by the undead look and feel like.

Rincón Gallardo's treatment of land defense through queer aesthetics like camp, irony, and frivolity makes an important and unusual intervention in antiextractivist art. As Nicole Seymour has pointed out, environmental movements have tended to eschew the flamboyant aesthetics of queer culture and have gravitated instead toward "the complete opposite sensibility," characterized by austerity, sacrifice, and self-seriousness.[64] While Seymour writes specifically about the United States, the same can be said for Mexican and other Latin American environmentalist art, which rarely recurs to the exaggerated registers and affective excess of camp.

In many ways, the purported incompatibility between irreverence, on the one hand, and environmentalism and land defense, on the other, is understandable. As the high rates of land defender murders index, the defense of territory is serious business. And yet irreverent modes of mourning can complicate and complement mainstream representational modes, by preventing violence from being sublimated into trauma or spectacle and by pivoting toward desire. In mobilizing desire and playfulness, *Una trilogía de cuevas* actualizes Eve Tuck's suggestion, mentioned in Chapter 1, to resist reducing disenfranchised communities to the experience of damage. Instead, *Una trilogía de cuevas* teases out the complexity of life lived in the "extractive zone," in which pain, rage, joy, and desire intermingle.[65] These affects animate land defense, revealing it to be a processual, collaborative project sustained by dreams for a better future and memories of a shared past. Tuck elaborates on desire's qualities: "Desire is involved with the *not yet* and, at times, the *not anymore*"; it has a "ghostly remnant quality . . . not contained to the body but still derived of the body. Desire is about longing, about a present that is enriched by both the past and the future."[66] Following Tuck, desire is both subjunctive and hauntological: it is the discovery of possibilities, the charting of diverse paths for future action, the embrace of uncertainties, and the excess that courses alongside and throughout the experience of the real.

In *Una trilogía de cuevas*, Rincón Gallardo responds to dispossession by throwing a party. Metonymic of performance as an ephemeral doing, the party provides a space of refuge for the dispossessed to gather around desires that can still come into fruition through their sustenance in the present. As she puts it in the trilogy's third chapter, *Resiliencia tlacuache*, "La fiesta es la vuelta de ese tiempo pasado que puede ser futuro, que habita en nuestros sueños del presente el regreso de esas fuerzas que le dan vida a uno." (The party is the return of a past time that can become future, one that inhabits our present dreams when the forces that give life to the world return.) The trilogy stages a series of parties around different events: Bety Cariño's journey to the underworld in *El viaje de formol*, the ghosts and toxic waste accumulated over centuries of mining in Zacatecas in *Sangre pesada*, and land defense in San José del Progreso in *Resiliencia tlacuache*.

Each episode in the trilogy is made up of multiple subchapters that adopt a wide range of expressive modes, including music videos, choreographed dance, dramatized scenes, and even video games. This miscellany jostles together, colligating a variety of seemingly dissimilar representational and affective modes, from the somber to the irate to the raucously sensual. Each episode is also centered around a nagual, or guiding spirit animal: the axolotl; the hummingbird; and the opossum, respectively. Personified by a costumed Rincón Gallardo, the nagual takes on the role of narrator, and its

FIGURE 2.8. The opossum (Naomi Rincón Gallardo) appears in *Resiliencia tlacuache* as Rosalinda Dionicio's nagual. *Resiliencia tlacuache*, Naomi Rincón Gallardo, 2019. Photo-documentation by Claudia López Terroso. Courtesy of the artist.

traits orient the episode's theme. The axolotl's ability to regenerate its limbs, the hummingbird's ability to migrate thousands of miles, and the opossum's ability to play dead embody survival strategies that parallel the experiences of the real-life activists who inspire each episode, articulating an expansive idea of territory defined by the confluence of human and nonhuman forces.

In *Resiliencia tlacuache*, the opossum mirrors Zapotec lawyer and antimining activist Rosalinda Dionicio's brush with death in 2012, when paramilitary forces linked to the aforementioned Minera Cuzcatlán and municipal authorities ambushed her vehicle. Bernardo Vásquez Sánchez was killed in the attack, but the bloodied Dionicio, who was shot twice, survived by pretending she too was dead. Because of this ability to endure attack, headstrong, resilient people from Dionicio's hometown of San José del Progreso are called *tlacuachitos* (little opossums), a detail that inspired Rincón Gallardo to conflate the two in her homage to Dionicio and other Zapotec land defenders.

At the same time that the opossum represents resilience and miraculous resurrection, it is also a trickster: the deity of drunkenness and thievery. In Mesoamerican myth, the opossum appears as the creature who brought humanity maize and fire, pilfering these life forces from the gods.[67] This association is enacted in *Resiliencia* through scenes that depict the personified opossum exuberantly singing, carousing, and offering tobacco. Dionicio does not appear on-screen; the choice to unfold her persona through the figure of the opossum avoids flattening her into the role of either idealized hero or passive victim. It instead positions her both as an activist who risks her life in defense of Zapotec territory and as a *pícara*, a rogue who relentlessly undermines her foes. The ludic representation of Dionicio as intoxicated trickster reorients the viewer's understanding of land defense as exceeding cliché associations with self-serious moralizing, rendering it compatible with pleasure, playfulness, and indulgence (fig. 2.8).

Rincón Gallardo's trilogy thus centers the violence experienced by land defenders but departs from the reverent representational modes analyzed earlier in this chapter. Whereas these modes idealize land defenders in an understandable attempt to foreground their heroism and innocence, drawing on nostalgia and revolutionary ideals to appeal to the widest possible public and drum up support to contest impunity, Rincón Gallardo cultivates more complex affective engagement with activists. The trilogy dramatizes land defense as outrageous fun. Its queer dance parties soaked in drink and song crack open a world outside of heteronormative and extractivist space-time, possible worlds that exemplify what Jill Dolan calls "utopian performatives," which "persuade us that beyond this 'now' of material oppression and unequal power relations lives a future that might be different, one whose potential we can feel as we're seared by the promise of a present that gestures toward a better later."[68] The affective excess that courses through the trilogy lifts the audience out of the present and into the imaginative space of the potential. In this sense, the trilogy refracts the way that land defenders are mourned in their communities, as leaders in the fight for another world whose deaths are not reducible to a narrative of victimhood, and makes them the impetus for renewed reimagining of other ways of living in a time of extractivist terror.

The trilogy's commingling of pain, intoxication, and pleasure is modeled after Latin America's *fiestas patronales*, or annual celebrations dedicated to different patron saints.[69] Festivities for these commemorative occasions often last several days to a week and are opportunities for the community to gather around religious ceremonies, feasting, drinking, singing, and dancing. Many elements of the trilogy map onto characteristics of the fiestas patronales: from the central presence of archetypal, gender-bending, and animal characters, to the use of masks, makeup, wigs, and handcrafted costumes, to the concluding fireworks that celebrate renewal through the destructive capabilities of fire. Drink and vice similarly play central roles in the second and third episodes, echoing Walter Benjamin's observation that intoxication, when critically appropriated, counteracts capitalism's numbing effects through the "loosening of the self."[70] Or, as it is playfully sung by the drunken opossum in *Resiliencia tlacuache*, personified by Rincón Gallardo in burlap costume: "en tiempos de despojo, que no haga falta el tepache" (in times of dispossession, may *tepache* [an intoxicating drink] not be scant).

The trilogy's analogy to patron saint festivals points to their similar construction as events that upend the status quo through improvised play. Anita Gonzalez observes in her study of festival dance in Oaxaca that the use of masks and nonhuman characters grants performers a sense of anonymity that frees them to break societal norms, escape into a world of corporeal expression, and act with impropriety. The performative embodiment of nonhuman characters, Gonzalez writes, transports "both performers and audience members into the realm of archetypal myth."[71] The characters in each trilogy episode bend time—as when murdered Mixtec activist Bety Cariño morphs into Aztec moon goddess Coyolxauhqui in *El viaje de formol*—creating a fantastical counterfactual order that renegotiates gender, sexuality, race, and politics.

EL VIAJE DE FORMOL: COUNTERFACTUAL MOURNING AS BENDING TIME AND SPACE

Una trilogía de cuevas originated in Rincón Gallardo's desire to honor and mourn Bety Cariño, the guiding figure of the first episode, *El viaje de formol*. In her artist's statement, Rincón Gallardo explains that *El viaje de formol* is "a twisted mythical critical fabulation aspiring to materialize and activate the ghost/spirit/body of the murdered Mixtec activist Bety Cariño . . . who dedicated her life to the opposition to extraction projects threatening the land, air, and water of Indigenous communities as well as their worldview and forms of social organization."[72] *El viaje de formol* follows Cariño's journey through the underworld, where she is cared for and cares for fellow female warriors, whom she trains in techniques of resistance, continuing to nurture the fight against dispossession even in death. In turn, these warriors help piece Cariño's dismembered body back together from where it floats in cosmic space in an act of feminist recomposition.

Cariño's speculative journey is framed by an axolotl narrator (performed by a costumed Rincón Gallardo), Cariño's nagual. Like Cariño, the axolotl is undead and exists out of time, the result of its preservation in formaldehyde by the German explorer Alexander von Humboldt, a character who appears in the episode's second subchapter. Standing in for the colonial gaze and the "personification of humanism," Humboldt falls in love with the axolotl.[73] The two share an erotically and racially charged encounter, in which Humboldt, performed as a drag king named Alex, is caressed from behind by Axol(otl) in an intimate embrace. Alex's desire for Axol subsequently morphs into possessiveness. He dissects Axol's body (now a handheld puppet), transforming his lover into an object of study, captured for perpetuity in formaldehyde. This tale of colonial longing and violent subjugation, the axolotl explains, is "another beginning of an endless story of appropriation," a frame narrative for contemporary extractivist dynamics.

As Cariño's nagual, or nonhuman coessence, the axolotl mirrors Cariño's experience. Both are brutally fragmented by colonial-extractive forces, yet, like the axolotl's ability to regenerate its limbs, in *El viaje de formol*, Cariño's broken body is pieced back together by a collective of women who cultivate a world in extractivism's cracks. Forcibly removed from time, they experience a state of suspension (akin to that of the titular formaldehyde) that is also a line of flight from the linear time of "progress." As ghosts, these undead companions inhabit a counterworld replete with speculative possibility, both in terms of its departure from the real and in terms of its unique transtemporal vantage, which weaves together continuities across colonial, patriarchal, and historical processes as well as across cosmic, human, and nonhuman forces. This excess seeps into and out of the songs, dances, orgies, and collectives that accompany Cariño's journey through the underworld, transmitting her "into the future in order to keep her undead."[74]

The futurity of creative practices undertaken in collective echo Cariño's own approach. In her posthumously published poem "Empujemos el silencio" (Push Back on Silence), Cariño concluded:

> No te venzas
> La poesía es resistencia
> Es otra forma de seguir vivas
> Pese a los gobiernos impuestos
>
> Don't give up
> Poetry is resistance
> It is another way of staying alive
> In spite of imposed governments[75]

Cariño's transtemporal configuration in *El viaje de formol* is cemented through her costume: a shiny polyester jumpsuit decked out with red tassels, plastic skull knee guards, a skull-laden belt, and blinking red nipples (fig. 2.9). Her vibrant handmade outfit echoes the Aztec moon goddess Coyolxauhqui's appearance in the enormous stone relief found in 1978 at the site of the Templo Mayor, the main temple, in what is now Mexico City. The

FIGURE 2.9. In *El viaje de formol*, the character of Bety Cariño (Bárbara Lázara) morphs into the Aztec goddess Coyolxauhqui. Naomi Rincón Gallado, *El viaje de formol*, 2017. Photo-documentation by Claudia López Terroso. Courtesy of the artist.

stone disk pictures a dismembered Coyolxauhqui after she was murdered by her younger brother, Huitzilopochtli, the Aztec war god, who sprang from their mother's womb upon hearing of Coyolxauhqui's plans to kill him. He tossed her body down the side of a mountain; her decapitated head became the moon.

According to Cecilia Klein, because all Aztec rulers were male, Coyolxauhqui was understood as a threat not only to her brother's power but "to the power and legitimacy of the state itself."[76] The disk depicting her broken body was placed at the spot in the temple where enemies were deposited after their ritual sacrifice. Klein notes that Coyolxauhqui's naked appearance on the relief was an intentionally humiliating presentation that marked her as sexually excessive and underscored her failed attempt to enter the masculine arena of politics. Coyolxauhqui's violent death immortalized in stone was emblematic of state power, "her female sexuality . . . a metaphor of the inferiority of all those who contested Aztec power, and of their inevitable political defeat."[77]

El viaje de formol emulates Coyolxauhqui's appearance on the relief. Though editing, Cariño's limbs appear severed from her torso; arranged in a dynamic circular composition, her body spirals as she makes her descent to the underworld. Rincón Gallardo's reactivation of Coyolxauhqui through Cariño articulates a transtemporal lineage of women who have been murdered for pursuing goals that flew in the face of patriarchy. It reclaims this figure who "represented the doomed or defeated challenger of male and state authority," repurposing her into an emblem of feminist resistance.[78] In this sense, Rincón Gallardo builds on the Chicana feminist tradition, including the work of Gloria Anzaldúa, who referred to Coyolxauhqui as a jumping-off point for the task of re-membering: interweaving the ancestral and historical past with personal stories of contemporary womanhood.[79] Similarly, Cherríe Moraga refered to Coyolxauhqui as a feminist icon who elided the reductive virgin/whore binary that circumscribed Chicana and Mexican women, embracing her as "la fuerza feminina, our attempt to pick up the fragments of our dismembered womanhood and reconstitute ourselves."[80] Rincón Gallardo similarly draws from the metaphors of dismembering and re-membering as narrative and representational ethos in which the task of art is reconstitutive, one of piecing back together. Given that Cariño died in an attempt to resupply the autonomous Triqui municipality of San José under paramilitary blockade (supported by the PRI), her collapse with Coyolxauhqui establishes a geneology of women who stand opposed to entrenched power. The invocation of pre-Columbian iconography here is not made in a gesture of nostalgic nationalism, but with a critical eye to the ways in which the state sustains power through violence, violence that is often wielded against women.

In *El viaje de formol*, Cariño is axolotl, Coyolxauhqui, and many more. In death, after she descends to the underworld with the help of widows who hold vigil, she multiplies, enacting the activist phrase "no murió, se multiplicó" (she didn't die, she multiplied). Neon lights illuminate her body as she collapses repeatedly to the ground, performing her corporeal disintegration. Superimposed against the distorted backdrop of the vigil's fire, as if floating in air, the undead Cariño character intones: "Soy muchas a la vez . . . no soy una no-no . . ." (I am many at once . . . I am not one no-no . . .) (fig. 2.10). She then lists the names of other women who have been killed for defending the land or protecting human rights, "Soy Bety, Bertha, Tere, Felicitas, Fidelia, Ester, Irma, Selena, No soy una no-no" (I am Bety, Bertha, Tere, Felicitas, Fidelia, Ester, Irma, Selena, I am not one no-no). With each name, a group of female warriors pop up around her in new configurations, striking a fighting pose. This multiplication of the undead sets the stage for the vaginal cave orgy to come, where many bodies come together in shared pleasure. This transition from multiplication to copulation highlights how the ethics of relationality that Cariño promotes has implications beyond activism. It is about finding pleasure in the company of others.

Performance scholar Laura G. Gutiérrez has pointed out that Rincón

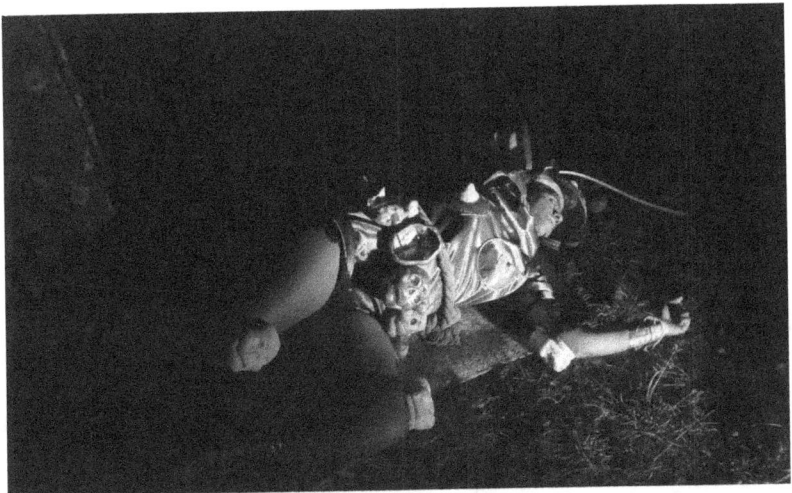

FIGURE 2.10. Bety Cariño (Bárbara Lázara) descends to the underworld in *El viaje de formol*. Naomi Rincón Gallado, *El viaje de formol*, 2017. Photo-documentation by Claudia López Terroso. Courtesy of the artist.

Gallardo's trilogy works in a parallel vein to Saidiya Hartman's "critical fabulation": it is storytelling that strains to tell impossible histories.[81] In her work about the lives of enslaved Black women, Hartman adopts this method because the historical record has obscured these stories. The impossibility of accessing them pushes Hartman to recur to subjunctive engagement with how their lives might have been, if she could fully imagine them.[82] Just as Hartman indexes the transtemporal possibilities of this exercise, which brings the past into focus without reproducing the language of violence, Rincón Gallardo draws out the reparative possibilities of encounters among women warriors across time.

This other space of the counterfactual possible is materialized in the trilogy's titular cave. Associated with the female body, sexual pleasure, and reproduction, as well as with the underworld and sleep, in each episode the cave is a central site of communion, refuge, and pleasure. In a pivotal sequence in *El viaje de formol* after Cariño's body has disintegrated into pure vibration, a raucous punk concert in the vaginal cave of the underworld transforms into a queer orgy. The heavy metal lyrics shouted by the revelers first in Spanish and then again in English make Rincón Gallardo's message clear: "The tomb doesn't stop us, we are not tired, neither disheartened, repeatedly dead, indomitable, illegible monsters, articulated Coyolxauhqui." The women rhythmically move their hands through the cave's muddy floor, sensually spreading mud on each other's legs and onto an enormous pink dildo strapped to one of the punk mourners. The orgy in the mud illustrates how other forms of politics are actualized through embodied desires that bring bodies into relation.

The titular cave is a transformative space where bodies can come into renewed contact. In Mesoamerican codices like the Mayan *Popol Vuh* and Nahua, Mixtec, and Tarasco accounts, the Seven Caves are where life originated. Chicomoztoc, the Nahuatl name for this primordial place of Seven Caves, or Tulan Zuyua, in K'iche' Maya, is an immanent space-out-of-time, a space of flux or transit between different times and worlds. Federico Navarrete explains that ritual allowed access to this alternate temporal and ontological order, in which there is no difference between humans, non-humans, and dieties. Those that emerged did so renewed, manifesting a new way of being and even a new historical era. "Era un borrón y cuenta nueva, o una declaración de independencia que iba mucho más allá de lo estríctamente político, pues afectaba la corporalidad misma de las personas, sus formas de vida y subsistencia, su identidad étnica." (It wiped the slate clean, or it was a declaration of independence that went far beyond the strictly political, since it affected peoples' corporality, their ways of life and of subsistence, their ethnic identity.)[83] Drawing on this conceptualization of the cave as a portal to another temporal-ontological space, a portal that is not just a temporary break from the status quo, but rather a transformative experience that engenders other worlds, Rincón Gallardo's *Trilogía de cuevas* deploys performance as a counterfactual feminist realm, inspired by land defenders like Bety Cariño and Rosalinda Dionicio.

MOURNING IN A COUNTERFACTUAL KEY

Writing about the conquest of the Americas, Adam Lifshey posits that the goal of empire was to suppress the many possible worlds associated with non-Western thought into one all-encompassing reality modeled after Europe's image, or what John Law has called the "one-world world," a world that presents itself as the only world possible.[84] We see this at work today in the way Minera Cuzcatlán presents itself as the only possible horizon for Oaxaca, consolidating the conceit of Oaxaca-as-mine. By contrast, the subjunctive, Lifshey elaborates, is "the antithesis of the declarative empire . . . where plural realities and possibilities exist side by side," the realm of "contestatory narratives, both those that already exist and those that could come into being."[85] Indigenous land defenders uphold subjunctive iterations of territory, or what Arturo Escobar describes as plural ways of making the world that reject the "design dream" of development that has organized life around consumption and growth.⁸⁶

Irmgard Emmelhainz points out that in a world organized around extractivist capitalism, lives that are normatively considered "good," characterized by a surplus of energy and consumable goods, are "premised on the fundamental act of violence: the destruction of the lives of others."⁸⁷ Just as commodities obscure their means of production, so too does extractivism obscure its infrastructure of violence. Attending to the violence inflicted on land defenders, who are disproportionately Indigenous, provides a necessary corrective to the lopsided focus on "slow violence" by ecocritical scholarship. Bare violence and slow violence go hand in hand as extractivist world breaking erodes historical configurations of territory to make room for Oaxaca-as-mine. While murder is just one of many forms of bare violence, it acquires particular resonance as the ultimate expression of power, a form of terror whose goal is the creation of "a world outside relation," in which those who are opposed to extractivism are exterminated or forced to "live at the edge of life . . . people for whom living means continually standing up to death."⁸⁸

If violence against land defenders is a strategy through which to forcibly impose extractivism as the sole possibility for relating to the land—a form of relating that forecloses others by contaminating water, eradicating habitats, and so on—counterfactual practices of mourning like the activist slogan "Samir vive" and Rincón Gallardo's trilogy refuse that foreclosure as definitive and affirm the futurity of postextractivist forms of inhabiting territory grounded in Indigenous practices of land defense. Counterfactual aesthetics of mourning notably eschew the evidentiary function so often yoked to works about impunity. In contrast with forensic works, like conceptual artist Teresa Margolles's use of materials derived from bodily remains to confront museumgoers with the viscera of necropolitics or the state ID photos of the forty-three disappeared normalista students repeatedly circulated in the wake of Ayotzinapa, evidentiary strategies are of less salience in the representation of murdered land defenders.

Indeed, the works under consideration here seem to reject altogether the mandate that the primary function of art is to bring these victims to justice. Not that justice is out of the question—one only has to look to the famous case of Berta Cáceres, a Lenca land defender who was murdered in Honduras for her opposition to the construction of a dam on the Río Gualcarque. Cáceres's case was finally brought to trial thanks to the relentless efforts of activists and artists who propelled her case to such heightened levels of national and international visibility that the government was forced to take action. Yet the staggering number of land defenders who are murdered every year in Mexico, combined with rampant impunity, means that the justice system is not always the horizon of interpellation. Rather, the counterfactual negation of death's conclusivity in slogans like "Samir vive" and Rincón Gallardo's speculative enactment of Bety Cariño's descent into the underworld are interventions that are not directed toward the justice system at all.

Rather, the counterfactual assertion of life in the face of death in slogans like "Samir vive" identifies that the stakes of public mourning lie not just in visibility, proof, or the pursuit of justice, but in envisioning a world that persists beyond extractivism: the counterfactual world where Samir Flores Soberanes still lives, a world sustained by those who keep his memory and desires alive. Similarly, the raucous and sensual parties enacted in Naomi Rincón Gallardo's *Trilogía de cuevas* create a space where those who are minoritized, excluded, exploited, murdered, or dispossessed can coalesce in a gesture that refuses death. The trilogy's ephemeral happenings constitute a refuge oriented around counterfactual desires encoded in phrases like "Samir vive," in which life and territory are valued outside of profit. This subjunctive potential of art to create spaces and forms of signification that imagine beyond what is echoes and extends the important work performed by land defenders, who treat the future as a territory to be defended.

CHAPTER 3
Extinction Poetics

In 2018, Banco de México, the country's central bank, began to roll out a new set of banknotes to commemorate the quincentennial of the fall of Tenochtitlan. The seventh set of bills released by the mint since it opened in 1969, the "G family" is organized around the theme "identidad histórica, patrimonio natural" (historical identity, natural patrimony). Each bill features a person or place representative of a key era in Mexican history on one side and, on the other, flora or fauna from one of its ecosystems. Many of the featured animals are flagship species: familiar animals with widespread appeal that serve as ambassadors for specific habitats. These include species considered vulnerable to endangerment, like the monarch butterfly and the jaguar, as well as species that were previously endangered but have since recovered, like the gray whale. The banknotes place renewed importance on biodiversity in the context of the ongoing era of "biological annihilation" otherwise known as the sixth mass extinction, in which nearly one million species are threatened with extinction.[1] They consecrate a national pantheon of nonhuman species equal in stature to figures like Sor Juana Inés de la Cruz, who appears on the 100-peso note opposite the monarch butterfly, or Benito Juárez, who appears opposite the gray whale on the 500-peso note.

Each banknote in the G family was carefully conceptualized, designed, and vetted. Focus groups reviewed design mock-ups. This resulted in interesting changes, as public documents about the process attest. The 200-peso bill dedicated to desert ecosystems was originally going to feature the Mexican gray wolf, which nearly went extinct in the 1970s and has since been the subject of semisuccessful conservation efforts. But focus group members found the gray wolf off-putting, noting that they thought of it as "traitorous" and "intimidating." They suggested replacing it with the golden eagle, a recommendation that was ultimately implemented.[2] Similarly, a subsequent focus group's dislike of the axolotl's "ugly" appearance resulted in design tweaks to highlight the amphibian's more "playful," "photogenic," and "friendly" traits.[3]

The decision to keep the axolotl was perspicacious. No other bill caused as much of an uproar online as the fifty-peso axolotl note when it was released in 2021. On Twitter, people celebrated getting their hands on the coveted bill and promised never to part with it. One Twitter user joked that the redesigned bill was the real reason the Mexican economy had stalled—not the ongoing COVID-19 pandemic—"porque la raza no quiere aflojar el billete de 50 pesos del ajolote" (because folks don't want to let go of their fifty-peso axolotl bill).[4]

Despite pressing the axolotl into the pockets of millions, the redesigned banknote does little to impress upon its beholders the axolotl's endangered status. Counter to its tranquil depiction in its native Xochimilco, the axolotl is nearly extinct in the wild because of agricultural runoff, pollution, and urbanization.[5] Just as the bill's depiction of Tenochtitlan on its obverse side can be read against the grain as dramatizing the state's symbolic celebration of its Indigenous past and simultaneous disregard of the political demands of Indigenous peoples in the present, a parallel abyss characterizes how the axolotl has proliferated in literature and culture—as a symbol for Mexican identity in Roger Bartra's *La jaula de la melancolía* (The cage of melancholy, 1987), alterity in Julio Cortázar's 1956 short story "Axolotl," and, as we saw in the last chapter, regeneration in Naomi Rincón Gallardo's *El viaje de formol*—and yet this visibility has not translated to habitat revitalization efforts or to greater regulation of development in Xochimilco. Visibility, then, can foster a sense of intimacy with endangered life, yet this visibility is contingent on anthropomorphic familiarity: the dissolution of the axolotl's weirdness into an idiom of cuteness. It also elides the drivers of biodiversity loss, confounding the commonplace that representation in and of itself is pedagogical.

A similar hypothesis is advanced by Mexican poet Isabel Zapata in her poem "Miembro fantasma" (Phantom limb). Included in her debut collection *Una ballena es un país* (A whale is a country), published in 2019, "Miembro fantasma" critiques the fetishization of extinct species. The poem collates a posthumous archive of the thylacine, also known as the Tasmanian tiger, tracing how its extinction was catalyzed by staged photographs that convinced settlers in Tasmania that the carnivorous marsupial posed a threat to livestock. The poem contrasts settler antagonism toward the living thylacine with the allure that the species acquired after its extinction in 1936. The fantasy that the thylacine continues to persist, Zapata suggests, is more digestible than the painful truth of its eradication. She concludes that the desire to deny extinction transmutes the extinct into a "miembro fantasma," or phantom limb, a presence that is felt even when it is lost, circulating as an idea detached from material reality. "Celebramos la vida que no existe. / La sombra que avanza sin un cuerpo" (We celebrate life that no longer exists. / The shadow that advances without a body), her poem contends, rather than valuing life while it is still here.[6] Human attachment to the extinct risks transforming it into a phantom limb, an anthropomorphic projection, a mirror or a metaphor that erodes the specificity of nonhuman worlds, or what Jean Baudrillard calls the simulacrum, which "masks the *absence* of a basic reality; it bears no relation to any reality whatever."[7] For this reason, "Miembro fantasma" suggests that the thylacine's true memory resides in the grass where it once stepped and not in the representational archive of anthropogenic cultural production. This premise questions the very enterprise of representing the extinct, embedding a self-reflexive critique within Zapata's poem.

So how, then, to write about extinct or endangered species without transforming them into a phantom limb? This chapter takes up this question in its assessment of contemporary Mexican extinction poetics, defined capaciously as poetry concerned with past, present, or future biodiversity loss—but limited here to poetry that directly addresses nonhuman species. I contend that practitioners of contemporary extinction poetics deviate from the familiarity prioritized by the visual language of extinction, which, as we have seen with the fifty-peso note, privileges anthropocentric recognition, approval, and aesthetic appeal. By contrast, like Isabel Zapata, the poets considered here, Karen Villeda, Xitlálitl "Sisi" Rodríguez Mendoza, and Maricela Guerrero, approach the act of representation with self-reflexive ambivalence. Their poetics of extinction problematize the distortional anthropocentrism of representation yet nonetheless persist in locating in poetry a form through which to reinvent attunement to endangered life. I argue that this is because poetry, more than any other artistic form, facilitates interspecies contiguity, or proximity that does not consolidate ontological difference into anthropocentric recognition.

The poem is a contracted space where a vast array of things can be brought into relation. It is a relational form that invites associations from the conceptual to the sonic to the incongruent. As Zapata notes, "el poema lo permite todo" (the poem permits everything): a poem is a place where anything goes and everything fits.[8] Expounding this point, David Farrier writes that "poetry can compress vast acreages of meaning into a small compass or perform the kind of bold linkages that it would take reams of academic argument to plot; it can widen the aperture of our gaze or deposit us on the brink of transformation."[9] Farrier draws attention to the scalar work performed by poetry, its ability to toggle across time, space, and species. Extinction, a sprawling historical process set in motion by a range of human and nonhuman actors and interactions, requires the sort of scalar work that poetry facilitates.

Mario Aquilina argues that more than any other form of literature, poetry is defined by contiguity: sequence, contact, proximity, and bordering. In making this claim, Aquilina refers to the importance in poetry of the distribution of the words on the page, the arrangement of space (including blank space), and the recurrence of certain sounds, images, or tropes.[10] Contiguity is also the operative logic behind many figures of speech, like metaphor, simile, synecdoche, metonymy, oxymoron, and alliteration. Through these mechanisms of association, poetry can establish a relationship of proximity that retains uncertainty and opacity. Here I follow Emmanuel Levinas, who observed that "the proximity of things is poetry," meaning that the material encounter with the world is poetry and, vice versa, that poetry produces material contact with the world through sound and sensation. Levinas writes that art establishes proximity through "the immediacy of the sensible."[11] But this contact, Levinas cautions, takes place under a dark light, meaning that it is not wholly legible.

Like rewriting and counterfactual mourning, I identify contiguity as another subjunctive aesthetic strategy that mobilizes imaginative engagement in response to foreclosure—in this case, extinction. Echoing how the subjunctive grammatical mood gives expression to imagined possibilities from a situated site of enunciation grounded in the real, contiguity puts readers in ephemeral contact with endangered nonhuman life, but only from within the realm of imagination. The hypothetical encounter with the nonhuman other in the space of the poem in contemporary Mexican extinction poetics never coalesces into the certainty of knowledge or the closure of understanding, but remains within the subjunctive realm of the potential, of proposition and suggestion, inviting interpretation and inquiry.

In addition to the formal importance of contiguity, and the self-reflexive, ambivalent approach to the act of representation, the third defining characteristic of contemporary extinction poetics is its structural formulation of extinction. This articulation echoes Patrick Wolfe's incisive description of settler colonialism as "a structure not an event."[12] By this Wolfe argues against thinking about colonialism as a period in time that is demarcated by an end date, when it should be understood as an ongoing process that structurally informs the present. Extinction, too, is often collapsed into a specific moment in time: the death of the last known individual of a species. This identification of extinction as an event translates to narrative modes that are retrospective or anticipatory, nostalgic or fatalistic. Yet such an approach obviates how extinction unfolds in the present as part and parcel of the capitalist commodification of the planet, and how past extinctions continue to resound in and produce the present, as the poets in this chapter make clear.

In what follows, I address how Karen Villeda's *Dodo* (2013), Xitlálitl Rodríguez Mendoza's *Jaws (Tiburón)* (2015), and Maricela Guerrero's *El sueño de toda célula* (*The Dream of Every Cell*, 2018) situate extinction within a broader diagnostic of the "capitalist world ecology" formulated by Jason W. Moore to describe capitalism's operation as "a matrix of human- and extra-human nature premised on endless commodification."[13] Their poetics of extinction eschew the didactic or evidentiary task of convincing readers why nonhuman lives are worth saving, instead interrogating the historical and material circumstances that give rise to extinction and the role of poetry in subjunctively reconceptualizing the relationship across species in the era of accelerating biodiversity commodification. In their work, poetry comes into view as a medium that facilitates through contiguity the partial identification with endangered life and serves as the staging ground for experimental forms of expression that elude the instrumentalist grammar of extractivist capitalism.

THE FORERUNNERS OF MEXICAN EXTINCTION POETICS

Extinction first became a topic of concern in Mexican poetry in the 1970s, alongside the rise of the international Save the Whales movement and President Luis Echeverría's push to reclaim Mexico's coastal and marine resources. The gray whale, which mates and calves in the warm lagoons of Mexico's Baja California peninsula, was a focal point. Although its killing had been internationally banned as early as 1937 and Baja had been declared off-limits to commercial whalers in 1946, by the mid-1960s, pressure to reopen Baja to whaling spurred renewed debate about the ethics and economics of whale harvesting. In the wake of international outcry, discussions of reopening subsided. The Echeverría administration calculated that conservation "was a way of earning positive political capital for Mexico," both internationally, in terms of establishing its environmentalist credibility, and nationally, as a move that doubled down on marine resource sovereignty.[14] In 1972, legislation established the first sanctuary for gray whales in Laguna Ojo de Liebre (then called Scammon's Lagoon); in 1979, President José López Portillo followed suit, designating a second lagoon, Laguna San Ignacio, to be a sanctuary for pregnant whales. In this context, two precursors of contemporary Mexican poetics of extinction emerged: José Emilio Pacheco and Homero Aridjis. Both articulated the gray whale as the iconic threatened figure of Mexican modernity, illustrating Ursula Heise's observation that extinction functions to rehearse histories of modernization and contest the downsides of development.[15]

In 1976, Pacheco published a book of poetry, *Islas a la deriva* (Islands adrift), which included a section titled "Especies en peligro (y otras víctimas)" (Endangered species [and other victims]). There Pacheco interpreted extinction in the broadest of senses, as the possible foreclosure of everything from whales and birds to the work of art and the mermaid. This constellation of endangered subjects signaled that for Pacheco, extinction was sign and symptom of broader societal ruin. His poem "Zopilote" (Vulture) concluded that the disappearance of the Californian condor, critically endangered at the time, would trigger a world awash in trash. In "Ballenas" (Whales), the whale emerged as a cherished relic of an imperiled past, callously slaughtered for consumables like "lipstic, jabón, aceite, / alimento de perros" (lipstick, soap, oil, / dog food).[16] In "Augurios" (Auguries), included in the collection *Desde entonces* (From now on, 1980), Pacheco linked the disappearance of birds from his backyard in Mexico City to air pollution. Or perhaps, he speculated, the birds had sensed the city's imminent demise and fled.[17] In these poems, endangered species were an omen of societal collapse, both allegory for and the material consequence of accelerating industrialization and urbanization.

Homero Aridjis likewise foregrounded extinction in his poetry and activism. In 1985, Aridjis founded the Grupo de los Cien (Group of 100), a coalition of one hundred Latin American intellectuals united around the goal of expanding environmental regulations. From their position of cultural authority, the coalition amplified environmentalist and land defense efforts. Among the causes they promoted were the protection of the monarch butterfly in 1986, the push to improve air quality in Mexico City through the "Hoy no circula" (Not Running Today) campaign in 1987, and the ban on the commercial sale of marine turtles in 1990.[18] The Grupo successfully lobbied President Miguel de la Madrid to create El Vizcaíno Biosphere Reserve (which included Laguna Ojo de Liebre and Laguna San Ignacio) in 1988.[19] For decades Aridjis advocated for conservation causes in hundreds of newspaper editorials, a legacy that has since been compiled in the book of essays *Noticias de la tierra* (News of the Earth, 2012).

In tandem with these public-facing efforts, Aridjis's poetry tackled extinction by highlighting flagship species like the whale and the imperial woodpecker.[20] In contrast to the hopeful persistence of his activism, Aridjis's poetry was fatalistic in outlook, saturated in restless melancholia akin to what Emily Apter terms "planetary dysphoria."[21] Aridjis recurrently invoked apocalyptic, biblical images that imagined the last members of the human race surrounded by silence and destruction. Poems like "Descreación"

(Uncreation, 1990) were proleptic elegies that presented extinction as the inevitable endgame of humanity's destructive impulse. Others expressed frustration with extinction's invisibility. "Extinción del pájaro carpintero imperial" (Extinction of the Imperial Woodpecker, 1998) juxtaposed the imperial woodpecker's silencing with the continuation of business as normal in the logging industry. His fatalist poetics placed blame for environmental degradation at the feet of a universal humanity, which he problematically characterized in a totalizing sweep. Aridjis's declensionist poetics predicted a ruinous future, in implicit contrast to the pastoral nostalgia he ascribed to his childhood in Michoacán.[22] Avoiding this future informed Aridjis's pedagogical environmentalist poetics.[23] This didactic orientation motivated Aridjis to embrace a wide range of literary genres—from editorials and poetry to children's books—to reach divergent publics.

EXTINCTION POETICS TODAY

Extinction poetry in the twenty-first century pivots away from the doom and gloom that saturated Pacheco and Aridjis's poetics. This turn is somewhat surprising given that scientific evidence of ongoing biological annihilation is now more robustly alarming than it was in the 1980s. Yet it evinces how environmentalism and poetry have transformed over the last several decades.

First, doom and gloom is no longer a rhetorical strategy proper to environmentalists. It has been co-opted by the state to reinforce the futility of preventative action, as evidenced by 2020 remarks by President Andrés Manuel López Obrador that the jaguar should not be a factor in considerations about the planned Tren Maya megaproject since its habitat has already suffered from human encroachment.[24] The megaproject will lay 900 miles of train track in southeastern Mexico to shuttle tourists from the resorts of Cancún and Tulum to less visited jungle areas. The most controversial section of the proposed route will run new track through the Calakmul Biosphere Reserve and the Maya Forest, the second largest remaining expanse of tropical forest in the Americas after the Amazon. The region is home to several Indigenous Maya groups, including the Tzeltal, Tzotzil, Chontal, and Ch'ol, as well as the Zoque, who have hotly contested the project's implementation.[25] In addition to the strain the tourist influx will put on Indigenous communities' water access, preliminary studies show that disruptions caused by the train will endanger at least 315 species, from the yellow-headed Amazon parrot to the Yucatán black howler monkey.[26] The jaguar, a near-threatened species, is emblematic of the danger the megaproject poses: experts project that the Tren Maya will fracture the expansive territories the big cat needs to hunt and reproduce.[27] This splintering would impede the diversification of the jaguar's genetic pool and endanger its survival. Dismissing these concerns, AMLO remarked that "hay sitios donde hay poquita selva" (there are many spots where little jungle is left), arguing that even if the train's construction were to pose a risk, the threat would be negligible, since the jaguar's habitat has already been fragmented.[28]

This fatalist logic relies on a sense of helplessness to justify inaction; it affirms that it is already too late, denying that something could still be done to prevent further habitat destruction. In a context in which fatalism is used by the state to justify the continuation of ecocidal megaprojects, the role of the poet is not to be a prophet of doom, but to make the case for life amid conditions of endangerment, as we will see in the works considered ahead.

Second, in contrast with the visibility and clout of El Grupo de los Cien in the 1980s, contemporary poets reach more limited audiences. Ignacio Sánchez Prado has observed that contemporary Mexican poets face an "epistemic crisis" in which poetry has "lost its epistemological privilege," or the philosophical and social authority it once held.[29] This is attributable to a shift in the dynamics of state funding, which in the twenty-first century is characterized by the availability of grants to fund writers to conceptualize and execute creative works and the comparative paucity of funds directed toward the distribution of these creative works. Alejandro Higashi neatly sums up the situation as a "market of many competitors in a country of few readers."[30] As a result, Higashi writes, poets gear their writing toward specialized publics: namely, other poets. And because of the sheer quantity of new publications combined with distribution difficulties, a common complaint is that even poets don't read each other. Higashi concludes from this that contemporary books of poetry are "alejados del compromiso social" (removed from social commitment).[31] This sweeping statement, published in 2014, has since been deflated by countless poets who have mobilized around current events, including Ayotzinapa, the #NiUnaMenos movement, and Indigenous struggles for autonomy, signaling a recent shift in Mexican poetry back toward political activism.

Yet with the exception of numerous Indigenous poets—like Mikeas Sánchez, a cofounder of the Movimiento Indígena del Pueblo Creyente Zoque en Defensa de la Vida y la Tierra (ZODEVITE, Indigenous Movement of the Zoque People Who Believe in Defense of Life and Earth), a group that has blocked mining concessions on Zoque land, and Pedro Uc Be, whose important essay *Resistencia del territorio maya frente al despojo* (Maya Territorial Resistance in the Face of Dispossession, 2021) decries the Tren Maya megaproject as an extinction-level event for human and nonhuman inhabitants of Yucatán—today there is not the same clear alliance between environmental activism and poetry as there was in the 1980s.³² We see little organized, sustained cultural opposition to the Tren Maya of the sort there has been for #NiUnaMenos or Ayotzinapa. Calls are beginning to emerge on this front, illustrated by Irmgard Emmelhainz's observation in *Toxic Loves, Impossible Futures*, published in 2022, that in the face of government policies "aligned with protecting capital rather than humans and the environment, . . . we need to run to stand alongside people defending their territories in order to create a Mexican Standing Rock in Yucatán and elsewhere, massively mobilizing to join and support these demands."³³

While contemporary extinction poetry is not activist in the sense of enacting direct intervention in local conservation efforts, it is political in its identification of the modes of production driving extinction. Rather than conceptualize poetry as a didactic and evidentiary tool of activist amplification (a role currently dominated by documentary film), poetry here serves a distinct function. Isabel Zapata explains that, in contrast with her experience working with conservation groups in the past,

> [la literatura] me permitió establecer vínculos de empatía más inmediatos y potentes que la academia o el activismo y me permitió también compartir mis inquietudes de manera más eficiente en círculos un poco más amplios. Más que ofrecer respuestas o datos, los libros me regalaron espacios a explorar, referencias y pistas para continuar una conversación que en la que ya había empezado a sentir que chocaba con pared.
>
> [literature] allowed me to establish more immediate bonds of empathy than academia or activism, and allowed me to share my uncertainties more efficiently with somewhat wider circles. More than offering answers or data, books gifted me with spaces of exploration, references and clues to continue a conversation that I had begun to feel was leading to a dead end.³⁴

Following Zapata, poetry offers a space of subjunctive experimentation where uncertainty and contradiction can flourish, an arena in which to complicate singular, totalizing narratives and embrace the representation of endangerment as an arena under dispute.

THE DODO AS PHANTOM LIMB

Karen Villeda's *Dodo* (2013) exemplifies the self-reflexive quality of contemporary extinction poetics. Her subject is the dodo bird, the first living animal acknowledged in writing to have been eradicated by humans. Its disappearance in the mid-seventeenth century was a pivotal event in the nascent historical awareness of the irreversible impact of settlers on the environments they traversed. The dodo was the first categorical biological loss, Thom van Dooren notes, to be narrated in a way that implicated humans "causally, perhaps emotionally, and certainly ethically."[35] Ever since, it has been a source of fascination. An ungainly, three-foot-tall, flightless pigeon, the dodo possessed a strangeness that contributed to its iconicity of a lost prehistoric past. Its extinction paradoxically elevated it to a level of otherwise unthinkable visibility, another example of how, in Zapata's words, "we celebrate life that does not exist."[36]

The human hand in the dodo's extinction was minimized by literary and artistic representations for centuries. Charles Hoge has found that in the eighteenth century, the dodo's extinction was seen as justified: a logical and inevitable outcome hastened by human visitors to Mauritius. These early posthumous accounts were devoid of the sentimentality that would later define elegies to the bird. The dodo was widely perceived as "a monster unfit for survival," whose strange appearance, supposed gluttony, and rapid eradication meant that its extinction was God's will.[37] This view was perpetuated by artistic renderings based on overstuffed taxidermies that exaggerated the dodo's bloated form. These efforts promoted a revisionary account of extinction that framed the victim of "invasive practices as a fundamentally doomed creature."[38] The taxonomic binomial *Didus ineptus*, coined in 1766 by Carl Linnaeus, cemented the bird's official classification as "inept." The final page of Villeda's *Dodo* concludes with this taxonomic entry, dramatizing how the violence of extinction is obscured by the scientific record.

In the nineteenth century, cultural approaches to the dodo shifted. It became an endearing figure in children's literature, memorably appearing in Lewis Carroll's *Alice's Adventures in Wonderland* (1865) as a fictional stand-in for the author: an anthropomorphized character with a speech impediment, human hands, and a walking cane. The publication of Carroll's novel coincided with the discovery of a partially fossilized dodo in Mauritius. This confluence prompted a rush of imaginative reconstructions. A blank slate upon which fantasies of lost life could be projected, the dodo's representational legacy evidences the paradox outlined by Zapata: emotional attachment to endangered animals combined with disinterest in the causes behind their disappearance.

Referencing and intervening in this corpus, Karen Villeda's *Dodo* pushes back against the desire to observe the extinct as it once was. Instead, to narrate the dodo, Villeda suggests, is to narrate empire, maritime expansion, and commerce—all of which the poem brings into contiguity. *Dodo* reenacts the Dutch arrival to the island of Mauritius in the Indian Ocean in the late sixteenth century.[39] There sailors encountered the dodo, which proved an easy food source given its inability to fly and its unfamiliarity with predators. Less than one hundred years after the arrival of the Dutch, the dodo was extinct. Playing with and subverting the totalizing drive of the epic, the paradigmatic genre of excursions to faraway lands brimming with odd beasts, Villeda articulates Dutch imperial expansionism as a chaotic endeavor, replete with action but little agency. In contrast too with centuries of representation focused on the dodo's morphological weirdness, Villeda refuses the reader's desire to see the bird through the text.

Divided into seven sections, *Dodo* is chronologically structured. It follows seven sailors as they embark on their voyage, reach the island of Mauritius, encounter the dodo, explode into interpersonal violence, hunt the bird, and depart. *Dodo*'s formal elements are rigorously conceptualized. Each of the seven sections of the book contains seven paragraph-length prose poems; each poem is composed of seven verses; most verses are made up of seven words. In addition to the formal rigor of a structure disciplined around the number seven—including seven human characters—each poem is individually presented on the page, chunking the work into digestible pieces. This presentation offsets Villeda's tendency toward challenging abstraction, manifest through the frequent absence of proper nouns, verbs, or identifiable subjects, signposts that typically facilitate the reader's ability to follow the text.

Each poem is narrated primarily from an omniscient third person viewpoint, which provides fragmented images of a scene. This account is interrupted by italicized lines narrated in the first person. These first-person speakers interject observations that enrich or complicate the omniscient retelling. It is unclear who is speaking in these italicized moments, giving the sensation of a murky chorus of voices in a dark room. The effect is a reinjection of uncertainty surrounding the events; it invites the reader to wonder who is recounting them, and to participate in the piecing together of the history that led up to the dodo's extinction.

Similarly disorienting, despite the poem's linear chronology, its narration is severed from temporal cues. *Dodo* is recounted exclusively in the present tense, immersing the reader in Villeda's sequential tableaus. The lack of temporal cues decisively differentiates the experience of reading *Dodo* from reading a history of the encounter, as it is impossible to decipher how much time has passed or who is speaking. This grammatical confinement to the present deviates from typical extinction narratives, which tend to narrate in the past (elegiac or nostalgic) or future tenses (anticipatory, prophetic). By contrast, Villeda's exclusive use of the present tense indexes that the grammar of capital unfolds in the now, with no sense of future consequences nor past accounting of what has been lost.

Despite its titular protagonism, the dodo in *Dodo* does not appear as an identifiable character, but as a fragmented body defined by the shape of its swollen torso and curved beak: "un cuerpo abombado, pico larguísimo en gancho."[40] Villeda's dodo is barely sketched in, outlined through the briefest of mentions of its plumage or naked head. It is never brought fully into view for the reader, accessible only through the trace, like the visual impression of its footprint, generated with Villeda's fingerprints, inserted halfway through the book. The decision to inscribe the titular subject through the poet's own digits suggests the impossibility of reconjuring the bird through language. A spectral presence that can no longer be known, the dodo is reduced to a gesture. Villeda's partial view rejects the anthropomorphic tradition that has transformed the dodo into a cheerful character, a reconstructed whole, a spectacle of prehistoric strangeness. Her approach reflects and refracts the partial view afforded by fossilized life and the taxonomic record, as well as the utilitarian perspective of the sailors who approached its body as a series of consumable parts.

Eschewing the familiar human/nonhuman, subject/object binary, *Dodo* treats the seven sailors similarly, relentlessly deconstructing them from identifiable characters into a collection of parts. The first mention of the men introduces them as prepositional objects: "siete barriles como pretexto para catorce brazos" (seven barrels, a pretext for fourteen arms).[41] This presentation reinforces their status as labor: pairs of arms in the service of empire rather than the enlightened subjects of reason associated with the epic.[42] Villeda replicates the commodification of labor power by referring to the sailors by their armpits, lips, or thumbs. This deindividualization is furthered by doing away with grammatical cues that clarify who is speaking or acting. The overall effect is an account of empire that subsumes historical actors into an economic machine put to work by arms and legs, appetites and carnal desires, illustrating Marx's famous observation, "We are not aware of this, nevertheless we do it," about the abstraction of labor into value.[43]

In its narration of settler extinction as a form of "domination without subject," to borrow Robert Kurz's formulation, *Dodo* maps a web of synecdochical associations in which most characters are referred to, not by name, but by their bodies.[44] This is exemplified in a poem in Part IV that narrates the sailors' first encounter with the dodo, before immediately killing it: "Catorce sobacos que sudan la gota gorda. Un cuerpo abombado, pico larguísimo en gancho. Una hinchazón de párpados y siete trompas.

Siete narices chatas y dos manos toscas. Un pulgar curvo. *Un cuerpo abombado, vertimos miel sobre él.* Moscas por moscas, docena de labios resecos" (Fourteen armpits that sweat buckets. A convex body, a long-hooked beak. A swelling of lids and seven snouts. Seven flat noses and two rough hands. A curved thumb. *A convex body, we pour honey over it.* Flies and more flies, a dozen parched lips).[45] The near-total absence of verbs crafts a still life of the inaugural encounter. Despite the lack of verbs, Villeda's use of fragmented lists accelerates the pace of reading. Alejandro Higashi has observed that the inventories that appear throughout *Dodo* "literalmente obligan a leer sin reposo y muy rápido dejan sin aliento" (literally oblige one to read quickly, without stopping, without taking a breath).[46] At the same time, the reader must resist this acceleration and slow down to determine who each part references through context clues: the Admiral's hands, the Redhead's thumb, the dodo's beak. Similarly, the use of ambiguous words like *trompas* (snouts) abstrusely indexes species; we deduce they are human because of their quantity. This strategy of metaphorical substitution that describes through parts effectively places bird, man, and island on equal discursive footing. Veering between subjects and across species, the narrative disorients the reader, making us aware of our desire to reinstate a hierarchy of order that parses subjects (humans) from objects (birds and land).

Another effect of the absence of identifying nouns is that it becomes difficult to perceive whether the dodo or the men are the target of the narrated violence, a blurring of species that purposefully confuses the dodo's vulnerability with that of the sailors. As the book progresses, violence grows in crescendo: the sailors kill the dodo at the Admiral's urging. The slaughter is recounted through a list of fragmented parts—a snapped neck, a bloody heart.[47] The dead body is refashioned into an adornment, the feathers into a wig and the intestine into a choker. The Admiral forces the crew to kiss the dead animal while waving their genitals, urinating on the island in declaration of ownership. The abject intensity of this episode signals the inextricability of extinction and colonization, acts of extreme violence wrapped in rituals of masculinity that fragment both the human and nonhuman into pieces. It also links violence against animals with aesthetics, showing how the dead animal body is repurposed into commodity, trophy, luxury, power.

Karen Villeda's *Dodo* weaves the dodo's reduction to a fragmented, lifeless body into the broader spectacle of Dutch territorial expansion, foregrounding the brutality of empire that is normally minimized in adventure narratives. As such, it forms part of the tradition identified by Ursula Heise that renders the dodo "a recurrent symbol of the destruction of nature wrought by the imperialist expansion of European modernity."[48] Yet, unlike works that sentimentalize or fabulate the dodo, Villeda denies readers the pleasure of seeing it as an integral body. We can only glimpse fragments rendered through the poet's fingerprints or discursive shards of its broken-down body. What is unveiled in *Dodo* is not the dodo, but the imperial desires and drives that bring about extinction—a system that takes on a life of its own, propelled forward in the present tense.

CONTIGUOUS RELATIONALITY IN XITLÁLITL RODRÍGUEZ MENDOZA'S *JAWS (TIBURÓN)*

Xitlálitl Rodríguez Mendoza's *Jaws (Tiburón)* furthers Villeda's approach to extinction as a structure that informs present-day commodification of human and nonhuman life in Mexico. The fifth book of poetry by the poet from Guadalajara, *Jaws* was awarded the Premio Nacional de Novela y Poesía Ignacio Manuel Altamirano in 2015. The pacing of *Jaws* is breathless, staccato, always moving—like a shark. Rodríguez tracks the titular Jaws in seemingly endless and contradictory directions, a rhizome of associations from the plastic beast in Steven Spielberg's classic film to the well-worn metaphor of the ruthless businessman. Her book draws from a vast cultural archive in which the shark's status as apex predator looms large, a ready-made source of anxiety for a beholder suddenly aware of their vulnerability before the titular beast. But as the speaker of *Jaws* reminds us, the shark always dies in the movies. Inverting the terms in which the shark is predator and the human its prey, *Jaws* insistently brings the shark back into view as a real, fleshy being, mobilizing endangerment as the basis of interspecies identification. In contrast with Villeda, who maintains a disciplined distance in *Dodo* between the reader and the titular bird, Rodríguez adopts a different tact, using poetry as a site to explore a shared identification with sharks on the basis of commodification. But the basis of this identification is never fully secured; it is possible, subjunctive, up in the air.

The speaker in *Jaws* drives forward in constant contradiction, back and forth, affirming in one breath that she is Jaws and in the next breath declaring Jaws to be an other. The catalyst for this ontological unfolding is the speaker's sleep apnea, a disorder in which breathing intermittently stops and restarts while one is sleeping. Momentary asphyxiation awakens the speaker, who is reborn during the dreamlike trance of interrupted sleep as Jaws: "tintorera / como todo animal tengo nombre / y mi nombre significa / representa / Jaws es otro pero en el fondo el mismo" (blue shark / like all animals I have a name / and my name means / represents / Jaws is another but, in the end, the same).⁴⁹ Enjambment establishes proximity and slippage between meaning and representation, between Jaws as the self and as the other, themes that structure the first section of the book, titled "Esto es agua" (This is water).

Attributed to David Foster Wallace in the epigraph, the phrase "Esto es agua" is reiterated such that it becomes a refrain. Water is the shared source of life: 60 percent of the adult human body is water; sharks extract oxygen from water in order to breathe. And yet the asserted identification between human self and shark self is constantly undermined by reminders of difference, like "asfixiarme es nacer" (to asphyxiate is to be born).⁵⁰ Which is to say that what is life ending for humans—submersion—is life-giving for sharks. Metaphorically, asphyxiation allows the speaker to transmute from one state to another: from death to life and from human to shark.

The poetic identification-disidentification with the shark maps a web of associations between human and nonhuman life. An organizing thread is the commodification of life on a planet organized by transnational trade. The poem "Aleteo de tiburones" (Shark finning) engages this theme by adopting the perspective of an asphyxiating shark whose fins have just been sliced off and who has been dumped back in the ocean to die. The practice of shark finning is ruthlessly efficient. Fishing boats optimize space by storing only the shark's most valued body part, the fin. The rest of the body, still alive, is thrown back to sea, where the shark sinks and suffocates, unable to continue swimming. One hundred million sharks are killed worldwide every year to meet the booming demand for shark fins, a traditional element in Chinese haute cuisine.

In "Aleteo de tiburones," the de-finned shark speaks to the reader as it sinks, its still-living body discarded in the waters of Mexico's exclusive economic zone, the two hundred nautical miles beyond the coastline where Mexico exercises sovereignty over marine resources. The de-finned shark's deathbed musings swirl together like the jumble of trash that clouds the water around it as it sinks. The breathless final stanza reckons with the denial that underpins the practice of releasing a creature back to the water after taking the parts it needs to live:

>Agua
>Saben que estoy vivo, que sigo vivo, que muero, que caigo
>No creen en mi extinción
>¡Soy increíble!
>Como el dolor dorsal

> Como el agua
> Sólo ella cree en mí
> Porque ha metido su mano en mis branquias
> Eso es.
>
> Water
> They know that I am alive, that I am still alive, that I am
> dying, that I am falling
> They don't believe in my extinction
> I am incredible!
> Like dorsal pain
> Like water
> Only water believes in me
> Because she has placed her hand inside my gills
> That's it.[51]

Here Rodríguez parses knowing (*saber*) about the harm caused by a practice like shark finning from believing (*creer*) that it leads to extinction. The only entity to which the speaker attributes belief in the shark's possible extinction is water, because of its proximity with the dying shark as it saturates its gills. Material proximity, then, is the means to believing in extinction, a form of understanding that goes beyond scientific knowledge. As humans this proximity eludes us; we cannot reach into the body of the dying shark. But poetry approximates this proximity by providing imaginative access to the de-finned shark's thoughts, a penetration that emulates the penetration of water in its gills. Rodríguez tempers this imaginative access by always circling back to contend with the representational baggage that clouds our understanding of the shark, encapsulated in the collection's title, *Jaws*, an allusion to the iconic blockbuster that cemented the shark's status in the collective imaginary as a villainous apex predator.

Overfishing has put a quarter of all shark species at risk of extinction. Shark depopulation has happened rapidly, their numbers dropping by 90 percent between the late 1980s and early 2000s.[52] Robust economic growth in China brought the shark fin, a status symbol and delicacy associated with health properties, into reach for countless consumers.[53] Mexico, one of the world's top ten shark-fishing nations, has seen similar declines in shark stocks. Of the 111 shark species that inhabit Mexican waters, many are considered collapsed (depleted by 90 percent) or overfished.[54] The shark-fishing industry continues nonetheless to be incredibly productive. Its catch rate peaked in the 1990s with an annual harvest of 34,000 tons, prompted by booming demand and the contemporaneous ban on sea turtle fisheries, which pushed fishermen to pivot to sharks. Over the past decade it has steadily maintained an annual catch rate of 25,000 tons.[55]

While the export of shark fins to Asia is the industry's primary driver, about 90 percent of the shark meat harvested in Mexican waters is consumed domestically, and Mexico regularly imports shark meat to meet domestic demand.[56] In contrast with the pricey shark fin, the relative cheapness of shark meat has made it an important food source for coastal communities. Mexico prohibited shark finning in 2007 and in 2011 enacted an annual moratorium on all shark fishing between the months of May and August. Despite these measures, sharks remain relatively unprotected; shark finning continues to be a common illicit practice, particularly by Chinese and Korean shipping boats that operate in international waters just outside of Mexico's exclusive economic zone.[57] Yet, as *Jaws* indicates, sharks are rarely seen as vulnerable to extinction—perhaps because of their lack of anthropomorphic features and their predatorial unlikability.

Through disjointed fragments and repetition, "Aleteo de tiburones" builds a sensation of vertigo that dramatizes the de-finned shark's drowning and its contiguous association with human precarity produced by neoliberal policies that have stripped away the social safety net. The discarded shark is akin to "a worker / [whose] contract [has been] rescinded two or three years before retirement." Shark embryos that eat their siblings before birth are aligned with Mexican children who must take computing and English classes as part of the "viviparous choreography of survival." These associations swirl together in a vertiginous bubble of thoughts:

> Acantilados de navíos no me salvan
> Nada salvo la asfixia
> Momento antes de morir
> Esto es agua
> Como al momento antes de nacer
> Coreografía vivípara de supervivencia
> Como clases de inglés y computación
> Voy cayendo
> Igual que un trabajador
> Contrato rescindido a dos o tres años de jubilarse
> Voy cayendo
>
> Cliffs of ships can't save me
> Nothing except asphyxiation
> Moments before dying
> This is water

> I eat in the moments before birth
> Viviparous choreography of survival
> Like classes of English and computing
> I'm falling
> Like a worker
> Contract rescinded two or three years before retirement
> I'm falling[58]

Rodríguez's linguistic play doubles as ontological play. The same words are repeated to different ends: *salvan* is invoked as the verb "save," then, in its prepositional form, *salvo*, as "except"; *como* is deployed to describe the act of eating and then for comparison. This unfolding of the same word in different directions of signification mirrors and reinforces the poem's analogical refractions across species, indexing a mutable being as well as metaphorical resonances across human and nonhuman life. In this manner, "Aleteo de tiburones" intersplices the structural commodification of life in Mexico into the singular event of a de-finned shark's death.

Successive poems further this work of bridging. "Aleteo de tiburones en tierra" (Shark finning on land) refashions the asphyxiation of finning into a metaphor for the experience of a man who looks out at the city from behind two inches of glass. Trapped in a corporate fishbowl, he wishes to be someone else, somewhere else:

> Ahora soy un hombre
> aunque quisiera ser actor, madre, o francés
> Quisiera ser algo que me alejara de este edificio
> de esta ventana
> de este vidrio de dos pulgadas que busca contener al vacío
> quizá ser extra en algún documental de National Geographic
> Quisiera no temerle a las alturas.
>
> Now I am a man
> although I would like to be an actor, a mother, or French
> I would like to be something that would take me away from
> this building
> from this window
> from this two-inch glass that tries to contain the abyss
> maybe I could be an extra in some National Geographic
> documentary
> I would like not to be afraid of heights.[59]

In stark contrast to the ecstatic porosity of the ocean, which allows for the imbrication of substances and things that was the only source of relief for the drowning shark, the corporate segregation of spaces and beings leaves the speaker no recourse but to dream in the subjunctive. The poem is a space where these aspirations can be voiced, their breathless concatenation emulating the porosity Rodríguez attributes to water.

The following poem, "Curriculum vitae," is narrated by Jaws, now an unemployed person who seeks work. Addressing a potential employer, the speaker details how they have moved like a shark from position to position, performing any and every task. In a panicked torrent, they reproduce the language of the market as if it were liturgy, affirming that water is a commodity: a substance for tea or coffee to increase the productivity of office workers, "blanca gasolina" (white gasoline). They detail how in a previous job they dumped truckloads of quality milk and cheese into the ocean to maintain the scarcity that buoys their cost, "porque de lo bueno poco / y como usted sabe, hay que devaluar / Señor" (because what is good is scarce / and as you know, we must devalue it / Sir), reproducing the language of capital to prove their value so that they might not be discarded like the de-finned shark.

The speaker of "Curriculum vitae" slips from the voice of the speaker seeking employment into that of the shark, both subjects who speak as commodities, rehearsing their work histories as objects that serve the tourist industry, who serve others, are hunted by others, and hunt others:

> También he servido
> en un hostal
> y repartido baguettes
> porque los tiburones somos útiles
> en el área del turismo
> pagan por vernos trabajar
> huir, digamos
> y tras nosotros funcionan barcos y tripulaciones
> redes en movimiento
> la cacería de mi cuerpo es el trabajo de alguien
> Señor.
>
> I have also served
> in a hostel
> and delivered baguettes
> because we sharks are useful
> in the area of tourism

> they pay to watch us work
> I mean flee
> and behind us the ships and crews labor
> nets in motion
> the hunting of my body is the work of someone
> Sir.[60]

The stream-of-conscious narration fluidly moves from the perspective of the unemployed man to that of the hunted shark by the sentence's close. These perspectives momentarily merge in the middle of the verse into a "we": "los tiburones somos útiles . . . pagan por vernos" (we sharks are useful . . . they pay to see us), a shared identification of the self as instrument and as spectacle, as a means of production and the product itself. Poetry's contiguous logic allows for a materialist conceptualization of the world that is fundamentally entangled yet ontologically diverse, as Rodríguez's poetic slippage between the voice of the shark and that of the contingent laborer illustrates.

This slipping, merging, and cleaving of human and shark speakers gestures at the shared political terrain that unites these distant life-forms—both preyed upon by capital—in a fleeting moment of commonality. This shared terrain dissolves at the start of the next poem, which closes the section "Esto es agua," as the speaker secures a job as a copyeditor: "pulcro y buen trabajo / manso trabajo para un pobre tiburón de camisa" (a neat and good job / a steady and docile job for a poor shirt-wearing shark). Adopting an exaggerated ingratiated tone to the unnamed "Señor" who has deigned to employ them, Jaws transforms into an editor who takes out their red pen, "esa mandíbula / y sancioné, marqué, juzgué el trabajo de los otros" (that jaw / and sanctioned, marked up, judged other's work), carrying out corrections to tame the language of others to meet the mandates of publishing houses. With this climactic wink, Rodríguez self-referentially situates cultural production as a site of commodified labor where meaning and representation are produced and the language of the market is reified.

Jaws' analogical ethos brings human and nonhuman into contiguous relation within the space of the poem, formally and thematically inscribing the role of proximity in engendering shared beliefs about the value of life beyond capitalist instrumentalization. In *Jaws* Rodríguez uses contiguity as a proxy for proximity to dramatize what poetry can do as an artistic practice beyond evidencing extinction or making it visible: it can serve as a method of expression that potentially reorients how we see, experience, and speak about the world. For another model that mobilizes contiguity to think relationally across species, I turn to Maricela Guerrero's *El sueño de toda célula.*

SUBJUNCTIVE DREAMS OF ANOTHER LANGUAGE

In Maricela Guerrero's *El sueño de toda célula*, thinking about planetary life originates in the self but moves relentlessly outward in recognition that the causes behind the sixth great extinction threaten, not just discrete forms of life, but the entire assemblage. Whereas Rodríguez plays with ontological indifferentiation between self and shark, Guerrero measures the physical distance between her body and a pack of Mexican grey wolves—557 kilometers east—and compares their bulk with that of her twelve-year-old son.[61] *El sueño de toda célula* writes the wolf's endangerment (hunted for its perceived threat to ranching, rendered precarious by habitat destruction in the name of real estate development) as homologous to the endangerment of human communities by femicide and dispossession.

Subtending these problems, Guerrero finds, is the "language of empire." Against that subtractive logic, *El sueño de toda célula* performs the search for another language, another way of communicating beyond the paradigm of endless, accelerating competition, summed up in the Olympic Games motto *Citius, Altius, Fortius* (Faster, Higher, Stronger).[62] Guiding Guerrero's search is a phrase imparted to her by her biology teacher, Maestra Olmedo: "el sueño de toda célula es devenir células" (the dream of every cell is to become more cells).[63] This pedagogy of care affirms the project of life to be world building, a pedagogy that Guerrero extends to poetry by asking readers to slow down, notice, yearn, and dream. The act of simply being—breathing, sitting, reading—is, for Guerrero, the first step in discovering the self to be entangled with "un río de lobos que alimente y limpie las palabras, las frases, las ideas imperiales que contra mis propios fluidos y linfas he pronunciado . . . un caudal y una lengua que acerque y fluya libre" (a river of wolves that might nourish and cleanse words, phrases, imperial ideas that I have uttered against my own fluids and lymphs . . . an aqueous flow and a language that might bring us close and that might course freely).[64] This yearning for another form of communication that has yet to fully materialize is articulated through the subjunctive, inscribing Guerrero's aspiration to speak to the world differently and refashion language as it might be.

Unlike the shark, the Mexican gray wolf went functionally extinct, and although it has since been the subject of semisuccessful reintroduction efforts, it is now considered the most endangered wolf subspecies in the world. *Canis lupus baileyi* is a genetically distinct type of gray wolf adapted to warmer climates. Its territory extends from the southern United States down to the Valley of Mexico. Proximate in size to a German shepherd, its coat carries the tawny tones of its habitat, as Guerrero notes: the creams, browns, and oranges of the mountainous sierra. Historically, as settlers moved deeper into wolf territory and overhunted small deer, wolves increasingly fed on domesticated animals and were seen as an existential threat to livestock. In the twentieth century, the United States carried out aggressive campaigns to eradicate the Mexican wolf. The state offered bounties that incentivized its eradication, and the US Bureau of Biological Survey (which later became the Fish and Wildlife Service) embarked on a campaign to trap, shoot, and poison it during the first half of the twentieth century. By 1970, the Mexican wolf population was wiped out north of the border. Yet at that same time, attitudes about wildlife began to shift. In tandem with the burgeoning environmental movement, the US Endangered Species Act of 1973 prompted an abrupt turnabout in policy that stressed the need to safeguard species from extinction.

Efforts to prevent the Mexican wolf's extinction in the 1970s involved tracking and trapping the last remaining individuals. Five were found in Mexico in the wild, and combined with three trapped animals, the US and Mexico initiated the Mexican Wolf Species Survival Plan with the goal of establishing a self-sustaining population. Animal sanctuaries and zoos on both sides of the border paired breeding partners and slowly expanded the captive population's genetic diversity. In 1998, the United States began reintroducing Mexican wolves to the wild in Arizona; in 2011, Mexico reintroduced five wolves to the Sierra Madre Occidental. The results of the latter experiment were disappointing: four of the five were hunted and the fifth recaptured. Reintroduction efforts nonetheless continued. In 2015, the first litter of pups was born in the wild in Mexico.

In her overview of these efforts, Pamela Maciel Cabañas, a biologist trained at Universidad Autónoma Metropolitana Iztapalapa who comanages the Wolf Haven International sanctuary in Washington State, explains the challenges facing attempts to reverse engineer decades of programmed eradication. These include the species' limited genetic diversity, the task of acclimating animals born in captivity to the wild, and, above all, the social acceptance of wolves by territorial competitors like coyotes, ranchers, and local residents. For this reason, Maciel Cabañas calls for greater funding to support efforts to "listen and work together with communities to explore how humans and large carnivores can share the landscape in sustainable ways"—that is, to do the cultural work of shifting predominant narratives about wolves that deem them "traitorous" and about territories as spaces that serve only human needs.[65]

Maciel Cabañas's description of the imperative to rewire the ways human communities relate to predators echoes what Guerrero describes as the work of poetry: to undo the language of empire and articulate other, more inclusive modes of relating across species. The Mexican gray wolf appears in *El sueño de toda célula* as an emblem of forging forward despite multitudinous counterforces. An image that appears repeatedly throughout the book, in various iterations, is of being carried on the she-wolf's back: "Vamos en el lomo de una loba bosque arriba" (We ride on the she-wolf's back up into the woods).[66] At times the wolf is carrying pups, at times the reader, at times the poet: "La loba me echa a su lomo y me lleva bosque arriba" (She thrusts me onto her back and carries me up the mountain).[67] The image condenses contradictory sensations of making an arduous ascent and of being safely conveyed. This inverts habitual conservation rhetoric that positions the human as the protector and engineer of endangered life. Guerrero's she-wolf is the agent that carries humans toward relational ethics, transporting the poet and reader toward another way to communicate outside the subtractive language of empire.

For Guerrero, the language of empire refers to the extractive paradigm that approaches the nonhuman world as something to be mined. Its operative verb is *sustraer*, whose meaning is organized around deduction: "to remove, subtract, take away, steal." This logic of dispersal, fragmentation, and accumulation is codified into law, legitimizing it and criminalizing anyone and anything that obstructs its flow. In the poem "Datos" (Data), Guerrero writes, "La lengua del imperio de nuestros días está cifrada en estadísticas, en ríos de datos fluyendo por redes de energía y siliconas, sales: que acumulan reglas y multas y cárcel a los que van en contra del imperio" (The language of empire of our time is coded in statistics, in rivers of data flowing through networks of energy and silicon, salts: that accumulate rules and fines and prison for those who go against it).[68]

This focus on the quantification and regulation of life describes not only the business of extraction but also the logic undergirding conservationism. Despite the Mexican gray wolf's close brush with extinction, it was only in 2015 that the subspecies was officially recognized by the US Fish and Wildlife Service as endangered. The law expanded the territory within which the reintroduced wolf packs could operate but restricted them from moving beyond Arizona's Interstate 40: an artificially imposed boundary. It permitted the population to grow to 325 members (from the 80 identified in 2015) but specified that if that ceiling were to be surpassed, excess individuals would be captured and shipped to Mexico.[69] Despite the gray wolf's relatively new endangered status, the law still allowed private citizens to kill gray wolves that posed an active threat to domestic and game animals. These regulations reflected lawmakers' unwillingness to think about how the encroachment of human settlers in wolf habitats might require humans and wolves to learn to coexist. Instead, the law framed the wolves' survival as something to be tightly controlled, perpetuating the combative framework that pushed wolves to extinction in the first place.

Against the zero-sum logic of extractive empire, *El sueño de toda célula* uses Maestra Olmedo's axiom, "El sueño de toda célula es devenir células," to articulate how life gives way to life in a process of endless becoming.[70] Guerrero's focus on becoming and the wolf pack resonates with Gilles Deleuze and Félix Guattari's *A Thousand Plateaus*, in which the wolf pack is a palliative to the constraints of bourgeois individualism. The pack signifies both a single collective and an irreducible multiplicity, a framework counter to the molecular logic of neoliberal capitalism. It points to a politics of becoming that lacks a true center, something that can be achieved, according to Deleuze and Guattari, through writing and dreaming.[71]

Guerrero concurs that writing and dreaming are means for moving toward a more material language, "un lenguaje hecho de manos y viento y nutrientes" (a language made of hands and wind and nutrients).[72] She finds this language in how plants and animals speak to one another, sharing information through the air and soil by releasing chemical compounds or other bodily fluids. Mycorrhizal fungi facilitate these networks, forming subterranean threads that connect individual plants and allow for the transfer of carbon, nitrogen, water, and minerals. The import of this information does not lie in its extractable encoded knowledge but in how it articulates a nonhierarchical way of speaking and breathing as an assemblage. The speaker recurs to the subjunctive mood to express her yearning to stretch her mind and tongue in a way that would allow her to become part of this nonhuman mesh, this reality that is there and yet is just out of reach: "Quisiera ser bacteria: una célula que se comunique en silencio. Quisiera comprender ese lenguaje de humus de nitrógeno de carbono de información fluyendo por las raíces de las secoyas las ceibas las casuarinas. Hablar en árbol. Hablar en lobo" (I would like to be bacteria: a cell that communicates in silence. I would like to comprehend the language of humus of nitrogen of carbon of information flowing through sequoia ceiba casuarina roots. To speak in tree. To speak in wolf).[73] The possibility of somehow altering human language or the mind to participate in this other way of communicating and relating is the subjunctive dream that organizes Guerrero's book. Instead of using realist techniques to detail how plants exchange information, Guerrero presents this information in ways that emulate its fluid transfer. The breathless list of elements and trees unbroken by punctuation imitates the horizontality among life-forms and the seamless way in which they exchange information. The lack of punctuation formally inscribes the aspiration to discard the anthropocentrism that governs thought; it also inscribes the power of desire to multiply and snowball, like cells that divide into other cells, to ultimately transmute the political and social status quo.

Guerrero articulates the yearning to speak otherwise, without words and through the porosity of bodies, as a form of shelter (*cobijo*) from the extractive drive to categorize and extract. It is this aspired-for refuge that Guerrero invokes with the line "quisiera saber si entramos todos" (I would like to know if we all enter).[74] The subjunctive unrealized desire for a space where everything enters, metonymic of the planet itself, composed of contiguous material assemblages, finds its expression in the body of the poem. This refuge, Guerrero posits, is what poetry can offer: a vehicle that transports the poet and the reader alike toward another way of being, even if it is not yet attainable. Like the speaker who is carried uphill on the she-wolf's back, *El sueño de toda célula* carries the reader toward forms of consciousness that are desired but not yet socially available, mobilizing literature's subjunctive capacity.

FROM SUPPLEMENT TO REFUGE

The phantom limb syndrome that Isabel Zapata critiques in *Una ballena es un país* signals a pitfall of human imaginative closeness with nonhuman species: the appreciation of symbolic value over embodied materiality. Zapata's poem also diagnoses the tendency to feel more invested in life-forms that are gone than in those that still exist. How then to represent endangered life without celebrating "la sombra que avanza sin un cuerpo" (the shadow that advances without a body)?[75] Scholars of animal studies have advanced a similar critique in their demand that the presence of animals in literature be read not just for their metaphorical import but as subjects with specific ways of knowing and being in the world.[76]

Yet, as Héctor Hoyos has suggested, literary figuration can sketch a bridge between the human self and the nonhuman world. "Metaphor, metonymy, allegory, and literary figures in general," Hoyos writes, "supplement that with which deduction and inference have difficulty grasping."[77] Hoyos's use of the word "supplement" to describe the function of figurative language is of particular interest, as it echoes the supplemental nature of Zapata's "phantom limb": an imagined presence that makes the body feel complete. These divergent invocations of the supplement point in opposing directions: one to imaginative enrichment, the other to misleading replacement.

Jacques Derrida's theory of supplementarity helps parse this ambiguity. For Derrida, the supplement is the thing that is added—representation—that serves a "cumulating function" in its enhancement of the thing it supplements. Yet in its additive fulfillment of that thing, the supplement also replaces the original: "it intervenes or insinuates itself *in-the-place of.*"[78] Representation as supplement is both an extension and a proxy that always points back to the thing itself, revealing the original's lack, for it is no longer complete without that extension. This, Derrida determines, "produces no relief."[79] This inextricability between the real and the figurative is further consecrated in extinction. Without an original, the supplement is all that remains.

Following this "undecidability" of representation as supplement to extinct or endangered life, the present chapter has examined how contemporary Mexican poets respond to this conundrum. Unlike the pioneers of extinction poetry in Mexico, contemporary extinction poetics practitioners pivot away from declensionist narratives and doom-and-gloom moralizing. Their emphasis lies less in how art can stand witness to the horror of the sixth extinction and more in probing literature's complicity in producing stories that obscure the causes behind extinction. Contemporary poets aspire to rework language and form to loosen the totalizing grip of anthropocentrism and the subtractive logic of extractivism.

Villeda, Rodríguez, and Guerrero mobilize contiguity to consider how the endangerment of human and nonhuman life is systematically produced by capitalist accumulation. Contiguity, as Eve Kosofsky Sedgwick has put it, manifests in a myriad of contradictory ways: "desiring, identifying, representing, repelling, paralleling, differentiating, rivaling, leaning, twisting, mimicking, withdrawing, attracting, aggressing, warping, and other relations."[80] In *Dodo*, *Jaws*, and *El sueño de toda célula*, contiguities swell and swirl together in a moving assemblage that gives shape to these jostling desires, identifications, and divergences between humans, dodo birds, sharks, and wolves. The poem becomes a refuge in which to experiment with as-of-yet unachievable desires, as Guerrero's use of the imperfect subjunctive underscores: "quisiera ser bacteria," "quisiera comprender ese lenguaje de humus" (I would like to be bacteria, I would like to understand the language of humus). Perhaps the yearning to transform contiguity into continuity is impossible, but the expression of this desire propels these poets forward, even as they are circumscribed by a damaged planet.

CHAPTER 4
The Rural Resilience Film

The current chapter considers recent documentary films that frame rural Mexico as a site under threat of foreclosure. Filmmakers point to the disproportionate impact of climate change on small-scale farmers dependent on rain-fed agriculture and draw attention to how erratic and reduced precipitation has destabilized traditional cultivation methods, spread food and water insecurity, and prompted new waves of rural-to-urban migration. Historically, Mexican cinema has treated the provincial interior as an unchanging site of escape, distant from the perils and promises of modernity. But in these films, affective cinematic investment in the rural undergirds anxieties about its tenuous future brought about by environmental dynamism. Filmmakers highlight rural inhabitants' engaged defense of their territories or, alternatively, depict the rural as a site of abandonment, ruin, and estrangement: a place emptied out by coalescing socioenvironmental forces. Of particular interest for the purposes of this chapter are films that focus on subjects who do not leave rural areas that others have deemed unlivable, whether by choice or for a lack of other options. I call this genre "the rural resilience film."

The films considered in this chapter, Everardo González's *Cuates de Australia* (Drought, 2013), Betzabé García's *Los reyes del pueblo que no existe* (Kings of nowhere, 2015), and Laura Herrero Garvín's *El Remolino* (The swirl, 2016), exemplify the rural resilience film. These place-based, observational documentaries capture sites of water scarcity or water excess, foregrounding subjects who refuse to accept the seeming end of their territories. Filmed respectively in Coahuila, Sinaloa, and Chiapas, each film adopts the modality of sensorial ethnography, eschewing didacticism to tease out the feel of communities that face the effects of climate change, development, and deforestation. Collectively they illustrate the rural resilience film's tendency to narrow the filmic scope to just a handful of profiled subjects, a frame that depicts the rural as largely emptied out, a space where those who remain are left to their own devices.

As a form, documentary might not seem to readily fit within the category of subjunctive aesthetics, given its long-standing association with facticity. Yet the subjunctive is well suited to explain the indeterminate way in which documentary images of environmental crisis are assembled by filmmakers and taken up by viewers. While the photographic image has traditionally been construed as a repository of fact and actuality, Barbie Zelizer elaborates that its meaning is also contingent. It is informed by the moment and circumstances in which the image was captured, but also by the way the image travels via viewers' imaginations to other contexts, allowing audiences to make "sense of the world in a way that is not necessarily rational, evidentiary, or reasoned."[1] Zelizer builds on Roland Barthes's conceptualization of *punctum*, or the image's "power of expansion" as it affects onlookers, triggering personal memories or an emotional response that lingers long after the act of seeing, ultimately shifting the image's signification.[2] Zelizer calls this "the subjunctive voice of the visual": the way in which an image gains meaning "beyond what it delineates and connotes at first glance" through its contingent consumption over time.[3]

Following Zelizer, I argue that the subjunctive voice of documentary film exists in tension with its engagement with reality and truth, allowing the film to speak in multiple ways and temporal registers that invite interpretation. The genre of sensorial ethnography, which unites the three rural resilience films considered in this chapter, is particularly rich terrain for the subjunctive voice of the visual because it privileges sensorial immersion over didactic explanation. The genre's typical lack of exposition allows the referentiality of the reality shown on-screen to idiosyncratically transform as it filters through viewers' memories or preexisting beliefs, tilting the documentary from an aesthetic form that reveals "what is" to one that sparks thoughts of "what could be." As such, the subjunctive voice of the visual is not inherently emancipatory, but ideologically ambiguous; footage of rural resilience can be taken up by urban film festival viewers, for instance, as indexing individual resourcefulness in the face of changing climates, rather than as an indictment of the state failure to ensure economic and societal stability, as we will see.

Attending to the subjunctive voice of the rural resilience film correspondingly requires subjunctive methodological questions that are not regularly employed by film scholars. As Zelizer puts it, the question is not necessarily what we see but what it "reminds [us] about or which possibilities it raises."[4]

With this in mind, in this chapter I pay particular attention to the triangulated relationship between film, filmmaker, and audience to investigate how the depictions of cataclysmic environmental events, like drought and flooding, in contemporary ecodocs (documentaries about environmental issues) are imbued through the subjunctive voice with nonfactual, affective, and speculative dimensions. The largely urban, upper-middle-class viewers of contemporary Mexican documentaries at domestic and international levels, I contend, consume films about rural crisis as speculative fictions. The unsustainable realities faced today by rural Mexicans interest these audiences because they validate their insecurities about what future climate change might look like while simultaneously affirming their distance from these realities.

RURAL RESILIENCE

Rural populations in Mexico are particularly vulnerable to climate change because of political restructuring that began in the 1990s. NAFTA accelerated the liberalization of the staple-food sector, slashed state support for sustenance farmers, and opened up domestic agriculture, changes that favored large agribusiness and monocropping.[5] These structural shifts depressed the price of corn and beans, weakened rural livelihoods, increased unemployment, and shrank rural residents' ability to withstand climatic stress.[6] Although climate change impacts urban dwellers—particularly in the form of heat waves—families dependent upon rain-fed agriculture and livestock cultivation disproportionately bear the brunt of the unfolding crisis. With the escalation of extreme weather patterns, the dry season is getting drier and the wet season wetter, leading to "extreme heat" and "historic water shortages," or to flooding that is catastrophic for crops.[7] Thus, while only 22 percent of Mexicans lived in rural areas as of 2010, the rural constitutes an outsized role in the cinematic environmental imagination.[8]

Statisticians have found that climate-driven changes to crop yield in Mexico are correlated to migration rates within Mexico and to the United States.[9] Predictive models suggest that intensifying extreme weather will prompt new waves of migration, particularly from coastal zones and areas sustained by rain-fed agriculture. The World Bank expects that migration will increase in the second half of the twenty-first century, noting that Mexico and Central America could produce 1.4 to 2.1 million climate refugees by 2050.[10] But this depends on people's ability to move. Under another scenario modeled by ProPublica that accounts for closed, militarized borders, rural populations are predicted to become less mobile and more vulnerable to hunger.[11]

Rather than presuppose migration as the only form of agency wielded by people in precarious environmental and socioeconomic circumstances, the rural resilience film shifts focus to those who decide to stay in place and the ways they make sense of the changing landscape. The rural resilience film can be understood, then, as the other side of the coin to the outsized focus on climate migration by international agencies dedicated to predicting the future impact of climate change as well as to the cinema of migration, a much more prolific genre.[12] Rural resilience stories are of interest to filmmakers and audiences alike for the lessons they impart about the interplay between global climate and local weather and between national politics and transnational trade agreements, and the stories they share about how to persist amid seemingly untenable circumstances.

Resilience is an ambiguous term. Scholars of Indigenous and postcolonial studies have used the concept to celebrate collective survival and flourishing in the wake of colonialism.[13] In this context, resilience describes a series of shared cultural practices and adaptation strategies that have allowed marginalized communities to weather repeated catastrophic disruptions. But resilience's conceptual stress on the individual ability to bounce back also aligns with the neoliberal abandonment of the state's responsibility to ensure collective well-being and the displacement of this responsibility onto individuals to come up with their own adaptive responses to crises. The implicit messaging of resilience is that people should accept situations of crisis and find ways to adjust, rather than seek to imagine how the state might operate otherwise.

In the context of climate change, Mark Vardy and Mick Smith suggest, the discourse of resilience has been mobilized to discipline subjects "into accepting radical precarity as if it were determined by nature alone."[14] The 2018 document about climate migration produced by the World Bank, *Groundswell: Preparing for Internal Climate Migration*, illustrates this point, with its subsumption of all of the accumulated sociopolitical vulnerabilities that shape hemispheric migration patterns under the term *climate*.[15] The rhetorical framing of climate crises as "normal 'regime shifts' to be adapted to," Vardy and Smith elaborate, diverts attention from how environmental changes could be combated through collective action and policy changes, such as "confronting the dominance of fossil fuel infrastructure and the corporate power of big oil."[16]

This chapter interrogates how the rural resilience film as a genre enters into these complicated and contradictory engagements with the concept of resilience. In its portrayal of practices of living, surviving, and even flourishing amid conditions of environmental damage, does the genre celebrate collective resilience, or does it reinforce neoliberal notions of the individual responsibility to adapt to change? The answer, I suggest, is both. The stories of those who have chosen to stay behind while much of their community has left demonstrate the human ability to adapt to seemingly impossible circumstances: life without water, life lived on water. However, by showcasing the resilience of rural Mexicans faced with collapsing ecosystems, some of these films paradoxically promote the idea that climate change can be adapted to without state intervention or systemic change.

Before turning to the cinematic representation of rural resilience, I first address the boom of ecodocumentaries in Mexico over the last decade and how the consolidation of urban elite audiences affects the genre's formulation and reception.

ECODOCS AND FESTIVAL DYNAMICS

Rocío González de Arce's exhaustive analysis indicates that more than 80 percent of all films ever made on environmental topics in Mexico were produced in the last decade, between 2009 and 2019. In the first decade of the twenty-first century, the annual rate of films with environmental themes (counted expansively, including shorts) hovered under ten. As of 2010, production dramatically jumped to an average of eighty films a year, indexing the consolidation of ecocinema as a genre.[17] This body of work is overwhelmingly dominated by the documentary. Although in recent years an increasing number of fiction features have begun to foreground environmental themes, such as the fight against mining that frames Federico Cecchetti's *El sueño de Maraʼakame* (The Dream of Maraʼakame, 2016), the black market for petroleum that is the basis of Edgar Nito's thriller *Huachicolero* (The Gasoline Thieves, 2019), the intersection of poppy cultivation, mining, and violence in Tatiana Huezo's *Noche de fuego* (Prayers for the Stolen, 2021), or the impact of irregular precipitation and disease on agave cultivation in *Dos estaciones* (Juan Pablo González, 2022), the documentary's minimal production costs and longstanding ties to activism make it the preferred filmic format for denouncing environmental degradation and otherwise examining environmental issues.

In addition to the genre's association with social critique, the boom of the ecodoc can be attributed to technological advances in film equipment that have made independent filmmaking relatively affordable, and the emergence of regional and international festivals dedicated to environmental topics. In Mexico, two environmentalist festivals have materialized in the past decade: Cinema Planeta, Festival Internacional de Cine y Medio Ambiente de México (FICMA MX), founded in 2009 as a spin-off of Spain's International Environmental Film Festival (FICMA), and ECOFILM Festival Internacional de Cortometrajes Ambientales, founded in 2011.[18] These specialized festivals, along with dozens of other film festivals that have popped up throughout Latin America and the world, expand opportunities for film distribution and funding avenues, augmenting the viability of creating films with environmental content that will actually circulate and be seen by viewers domestically and abroad.[19]

Documentary filmmaker Everardo González noted in an interview that the proliferation of film festivals was highly motivating for Mexican documentary makers, who now have access to wider audiences than was previously imaginable.[20] He went on to explain that he is cognizant of his audience's preferences, observing that international audiences are more interested in immersive imagery than in reading subtitles and would rather be emotionally moved than didactically taught. Referring to his documentary *Cuates de Australia* (Drought, 2011), which tracks an annual exodus made by residents of an eponymous ejido in Coahuila in response to drought, González conjectured that its visual focus on landscape was particularly appealing to transnational audiences:

> Es liberador ver películas con espacios abiertos, en los que la pantalla permite respirar sin llegar a ser contemplativo [. . .] Creo que por eso *Cuates de Australia* tuvo potencial internacional, pues cada vez más al resto del mundo le importa menos escuchar el español y leer los subtítulos; no interesa mucho no entender qué se dice, porque la imagen se cuenta sola.
>
> It is liberating to see movies with open spaces, in which the screen allows room to breathe without being totally contemplative. . . . I think for that reason *Drought* had international potential, because increasingly the rest of the world doesn't want to hear Spanish and read subtitles; they aren't really interested in understanding what they have to say, because the image tells the story itself.[21]

González attributes the international success of *Cuates*—perhaps implicitly in contrast to his dialogue-driven documentaries, *Los ladrones viejos: Las leyendas del Artegio* (The Old Thieves: The Legends of Artegio, 2007) and *El cielo abierto* (The Open Sky, 2011)—to its embrace of image over discourse and sensorial immersion over archival footage.

González's comment underscores that the cinematic focus on landscape and rural space is a valued commodity on the festival circuit. It has been so, Jens Andermann explains, since the earliest moments of Latin American cinema, which "sought to cater to the demands of vernacular as well as foreign audiences for recognizably 'national' locations in order to . . . add new territories to the expanding geography of the filmic image."[22] Audience engagement with far-off, unusual landscapes, like the arid Coahuila desert in *Cuates*, is cultivated through framing, editing, and the use of shots that linger on the unusual beauty of this hostile terrain.[23] This provides viewers with an immersive experience of alterity that does not demand too much from them—unlike the "dreaded" subtitles of González's earlier films. While scholars like Andermann have highlighted a tradition of Latin American film that resists and counteracts the "neocolonial and subalternizing geopolitical inscription" of rural landscapes by refusing to provide the "gaze any sure footing," it is nonetheless important to temper such assertions by recognizing that art house films circulate within global film markets in which viewers continue to find comfort and value in the aesthetic experience of landscapes, even when those films engage with rural landscapes "for their intrinsic properties, as localities," rather than for their symbolic value.[24] González's nod to the ease with which international viewers consume rural landscapes indicates that contemporary filmmakers continue to grapple with the marketable allure of local color and rural nostalgia first popularized by *costumbrismo*, a nineteenth-century literary and pictorial genre that paid attention to the details of everyday life and indulged in romanticized depictions of folkloric customs. This allure also holds for domestic viewers, as evidenced by the glossy coffee table book of black-and-white stills from *Cuates* published as part of the film's multimodal marketing of hostile rural aesthetics to middle- and upper-class consumers.[25]

Documentaries like *Cuates* that foreground natural phenomena like drought are also a valued commodity on the international festival circuit because they respond to the growing global appetite for films that treat climate change as a shared universal concern—albeit one whose impacts play out differently and unevenly across the globe. This maps onto Tamara Falicov's argument that festivals have privileged and consequently perpetuated a globalized art house aesthetic that is local in setting but universal in theme, an assertion that is confirmed by González's remaks about international viewers' preference for images of rural Mexican subjects over listening to what they have to say.[26] Falicov builds on Miriam Ross's analysis of the unintentional effects of a system in which Global North funders and viewers consume products made by and/or with Global South subjects. Ross writes that this "uneven benefactor-beneficiary relationship" furthers certain representational tropes, such as the foregrounding of national setting, poverty, and marginalized subgroups. Filmmakers are aware, Ross argues, of the international audience's implicit expectations and desire to consume "authentic" depictions of the developing world.[27] Delivering authenticity (images of drought in a stunning landscape) with little dialogue (minimizing the labor of reading subtitles) is therefore the task set out for the Mexican filmmaker who wishes to reach a wide transnational audience. These dynamics also play out at the domestic level. Ignacio Sánchez Prado explains that audience demographics for domestic art house cinema have coalesced in the twenty-first century around the middle to upper classes, resulting "in the rise of documentaries that advance the cultural and political values of the social elite," films that are purposefully "semiotically open ended" and ideologically ambiguous to appeal to a wide array of viewers.[28]

The demographic composition of national and international audiences of Mexican ecodocumentaries is crucial to understanding their subjunctive reception. Most scholars who have written about Latin American environmental cinema have focused their analysis on the films' content or activist components rather than on the spatial, racial, and classed elements of their production and reception. These methodologies frame the documentary as a neutral conduit: a mode of capturing concrete environmental issues and circulating them to wider audiences. Roberto Forns-Broggi's pioneering assessment of the Latin American ecodoc, for example, posits that cinema illustrates, documents, and communicates ideas that resist absorption by the market, such as the anticapitalist philosophy of *buen vivir* (or *sumak kawsay* in Quechua), which articulates individual well-being as inextricable from the well-being of the community and the environment. Ecocinema, Forns-Broggi argues, "serves to reinforce activist efforts" by bringing marginalized knowledge to the mainstream.[29] Yet, as Jorge Marcone has pointed out, representations of environmental struggles led by Indigenous or other marginalized groups are often packaged in terms that are accessible to Western viewers and fail to adequately represent the more-than-human cosmologies that drive them.[30]

Building off Marcone's critique, rather than see the documentary as a neutral conduit that resists market absorption, through the concept of the subjunctive voice of the image we can think about how the relational dynamics between filmmaker intent and audience expectations shape the filmic product and the meaning it accrues, which allows for a more nuanced take on the potentials and pitfalls of rural resilience films. The educated, upper-middle-class viewers who constitute the primary demographic of festival publics can safely be assumed to already have an investment in environmental issues. We might conjecture that such an audience expects ecodocs to validate their existing beliefs and anxieties about climate change. By watching films about how climate change plays out in the present in rural communities, such viewers might engage with that lived reality speculatively, as a preview of the future that lies in store for the world as the impacts of climate change become more widespread and a possible road map for how to survive or adapt to environmental changes.[31] While this hypothesis is itself necessarily speculative, the interpretive leap allows us to consider how the rural resilience film activates apocalyptic imaginaries of climate crisis and simultaneously sublimates the severity of these crises through tales of adaptation, such that "'resilience' becomes a way of assessing the post-disaster situation sociologically without treating it politically."[32] Put differently, images and stories of poorer rural folk adapting to ongoing climate change transform rural resilience into a commodity for the consumption of urban viewers whose fears about climate change's potential to rewrite the nation and the world—through migration, unemployment, food scarcity, poverty, and other crises—are both confirmed and ameliorated by stories of individual survival.

This is not to argue that documentarians thinking proactively about how their films might reach an international audience or reflecting retrospectively as to why they were successful in doing so is somehow a discredit to their work. Instead I aim to underscore that Mexican ecodocs are shaped in tandem by the director's vision, the subjects they depict, and viewers' expectations. In *Cuates de Australia*, careful cinematographic attention to the arid landscape through medium and long shots that immerse the viewer in the space draws in transnational audiences, but also reflects González's pointed rejection of the genre's long-standing humanitarian ethic, which typically endeavors to "restore" the humanity of a marginalized group living in conditions of environmental crisis by giving them a voice. This voice is habitually reinforced through pathos, meaning that the audience is compelled to feel an ethical obligation to the disenfranchised other through the sight of their suffering.[33] Advancing a line of critique that denounces the commodification of suffering, González has argued that ecocinema holds viewers emotionally hostage, making them feel pity for victimized subjects shown on-screen or guilt about their own relative privilege.[34] Because of this, he unequivocally rejects the term *ecofilm* as a descriptor for *Cuates*, arguing that his documentary has no activist content or environmentalist goal.[35]

Similarly, both *Los reyes del pueblo que no existe* and *El Remolino*, which were also successful on the festival circuit, privilege imagery, ambiance, and affect over didactic explanations and overt activism, signaling a trend in the Mexican ecodoc away from explicit activist messaging and toward aesthetic/

sensorial immersion. While some iterations of the genre continue to formulate politicized critique in the vein of the social documentary—as in the case of a film like Alberto Cortés's *El maíz en tiempos de guerra* (Maize in Times of War, 2016), which frames Indigenous efforts to protect seed diversity as the legacy of the Zapatista movement—the films that are most successful on the national and international festival scene tend toward a more subtle or diffuse communication of the politics of environmental change, and they privilege aesthetics as the means through which to encourage viewers to attend to rural spaces in slow-moving ecological crisis.[36] This move away from the evidentiary and the didactic in the ecodoc illustrates this book's overall argument about the ascendence of subjunctive aesthetic modes that embrace uncertainty and speculation in Mexican cultural forms that deal with issues like climate change and environmental crisis. Yet as this chapter demonstrates, the turn away from pedagogical filmmaking has mixed results. On the one hand, rather than tell the audience how to interpret the scene on-screen, these films cultivate the audience's attention to the rhythms of life in hostile environs through sensorial immersion. But on the other, the lack of exposition also allows footage of rural resilience to be taken up subjunctively in ways that potentially erode its truth value, for instance, by refashioning irregular precipitation into a coherent narrative of drought.

CUATES DE AUSTRALIA: REORGANIZING THE TIMESCALES OF DROUGHT

Everardo González's abnegation of the term *ecofilm* for its associations with activism and pathos is a generative point of departure for an extended discussion of his *Cuates de Australia*, which exemplifies what I call the rural resilience film. *Cuates* is a portrait of the residents of Cuates de Australia, an isolated ejido (communal land) in Coahuila that has no electricity or running water. Its inhabitants, ranchers who raise cattle, tame horses, and cultivate maguey, habitually migrate in response to drought, leaving Cuates when it runs out of water. The destination of this exodus is not shown on-screen; instead, after the ranchers depart, the camera lingers to witness the drought's climax—observing the empty village and animals dying of thirst—until the cycle is resolved with a heavy rainstorm, which brings human and nonhuman life back to the ejido.

Intended as a meditation on the commingling of life and death, *Cuates* rebukes the notion that arid desert regions are empty or devoid of life. It presents the northern Mexican desert as a territory defined by the coexistence of beauty and hostility, joy and pain. The emotional texture of this aesthetic of hardship is sonically imparted through a haunting soundtrack of *cantos cardenches*, a melancholic regional a cappella genre named after the cardenche cactus.[37] Like in González's other works, the film's subjects exist at the margins of Mexican society but are not portrayed as victims. Despite the ejido's harsh climatic conditions and minimal infrastructure (which indexes an absence of state support), its inhabitants lead lives replete with dancing, working, fighting, and other forms of community making. Their decision to return again and again to Cuates communicates an uplifting message about human adaptation. It contests the idea that climate-driven migration is something new, since humans have long migrated in response to climate variability. It signals that migration does not necessarily constitute an end, a beginning, or even a single trajectory, but can be reactive, cyclical, and seasonal.

González is part of a cohort of male filmmakers, along with Nicolás Echevarría, Juan Carlos Rulfo, and Eugenio Polgovsky, cited as revitalizing the Mexican documentary at the turn of the twenty-first century by revisiting anthropological and rural themes with renewed cinematographic attention.[38] In contrast with traditional modes of ethnographic filmmaking that present unfamiliar cultures to viewers through expository image and narrative, González practices what scholars have called "sensory ethnography," a form of nonfiction that imparts an affective experience of what it is like to be with that group, in that place, at that time.[39] This immersive sense of place is achieved in *Cuates* through sound design that interweaves storytelling and interviews with the melancholic cardenche music and manipulated nonhuman sounds (buzzing flies, cicada whines, coyote yips), as well as through images enhancing the sensation that the audience is accompanying the filmmaker as he discovers the landscape in real time, like handheld nighttime shots of animal corpses illuminated by a diegetic flashlight. This layered sensorial approach imparts a sense of how drought is experienced and understood; it plunges viewers into what Andermann has called the "trance" of ecological engagement while crafting a story of resilience that is simultaneously hyperlocal and capable of being universalized.[40] With its long takes and minimalist narrative, *Cuates* adopts recent trends in slow cinema, a mode of filmmaking that "downplays event in favor of mood, evocativeness, and an intensified sense of temporality."[41]

The sensorial immersion in the austere temporality of drought has the effect of communicating not only a sense of environmental time but also nostalgia for a time removed from the sociohistorical context of contemporary Mexico. The relative absence of historical and spatial cues in *Cuates* presents the ejido as existing outside of the time and space of the neoliberal nation-state. There are few signs of "modern" life (no electricity, television, internet, paved roads, or plumbing); no political references; no mention made of organized crime or impunity. In part this is because Cuates is, in fact, distanced from these forces. Its hostile terrain is inhospitable to poppy production, sparing the ejido from paramilitary influence. Cuates's taxing cycle of drought is therefore what makes it a haven: a blend of dystopian and utopian conditions that operates as a potent imaginative space for the viewer aware of the violence that lurks outside the frame.

Cuates taps into nostalgia for a time when northeastern Mexico was defined solely by the hardworking virility of cowboy culture, rather than the politics of violence that Sayak Valencia has termed "gore capitalism" and the concomitant erosion of civil society and forced migration.[42] This nostalgia is gendered: the profiled subjects in *Cuates* are overwhelmingly male. Women appear circumscribed to secondary or maternal roles. In this way *Cuates* draws on long-standing national imaginaries that elevate the *charro* as the epitome of Mexican masculinity, celebrating this subject as hardy enough to successfully operate outside the state and seemingly outside of history itself. González has explained that this is intentional: a purposeful effort to paint a different portrait of northeastern Mexico that counters the prevalent narrative of a "hyper violent, bloody, militarized, and criminal" region by tracing the continuity of other forms of community and masculinity.[43] Yet this celebration of the cowboy is nonetheless stylized. As Tomás Crowder-Taraborrelli's review notes, the film gives "rural labor a certain glamour."[44]

Through the absence of historical markers and references to current events, González abstracts *Cuates* from human time—history understood in terms of years or political events. Its plot is instead structured by environmental time: drought is the event that organizes the film's emotional and narrative arc. González presents drought as a linear and cyclical process. The film begins with scenes that establish normality—children attending school, ranchers breeding horses, animals drinking from the lake, people collecting water at the reservoir. As the film progresses, water begins to run out. The lake dries up and becomes thick mud; the reservoir shrinks (fig. 4.1). This culminates in the nadir of desiccation: where the lake once existed, there is dry and cracked earth; animals die; vultures circle. With the total absence of water, the villagers begin their exodus. At this climax, the camera does not follow the townspeople to their next destination but lingers in Cuates' desolate, abandoned landscape. The film concludes with the arrival of a torrential rain that ends the drought, refills the lake, and brings back the inhabitants, implying the cycle will recommence.

The denouement of heavy rain coincides with the birth of a child. This pregnancy is tracked throughout the film: the time frame of gestation

FIGURE 4.1. The water hole's expansion and contraction in *Cuates de Australia* (2011) is a visual device used to illustrate the timescales of drought.

overlaid onto that of drought. The unnamed mother becomes pregnant at the film's outset; the fetus is declared malnourished coincident with the drought's climax; and the baby is born in a sequence intercut with shots of lightning that signal the end of drought. The correspondence of pregnancy with drought proposes that waiting for rain is analogous to the anticipation of a child and that drought is a period of gestation filled with hope and fear. But another consequence of this parallel is temporal: it implies that drought unfolds within a consecutive nine-month period. This was not in fact the case, since *Cuates* was shot over a period of three years.

González has explained in interviews that the chronology presented on-screen diverges from the captured footage. The nadir of drought that takes place at the film's climax was actually filmed during the first year of shooting, while the massive rain that coincides with the narrative denouement was filmed in the third.[45] This means that the initial two-thirds of the film, which represent normalcy, were shot in between the events that represent death (desiccation) and subsequent rebirth (rain), events that occur on-screen in immediate succession. The film structures the arc of drought as a progressive drying that culminates in social and material death—death that is suddenly broken by the arrival of heavy rain. But the narrative unfolding of drought as it occurred in real time was not so simple or even declensionist. During filming, total desiccation did not dramatically give way to torrential rebirth. Instead, small, sporadic amounts of rain brought about a livable normality. Amid this sporadic precipitation, a massive rainstorm randomly occurred: the first storm large enough in three years to refill the lake and impactful enough for González to use to close the film. The film's reorganization, which poetically signals that death is followed by rebirth, is an artifice, one that possesses an easy logic: water diminishes until it is replenished. This contrasts with the actual progression of drought, in which rainfall is irregular in both quantity and sequence and does not follow any narrative logic in which climax gives way to resolution.

Does this artifice matter? González responds that the documentary's association with objectivity has long been and should continue to be problematized. He does not "want people to believe that what they are seeing is the truth . . . sometimes it's good to let people know they're watching a film."[46] González did not intend for *Cuates* to be a factual record of drought but a lyrical exploration of the experience of hoping for rain and adapting to its lack. This is why he affirms that *Cuates* "no aborda la sequía, sino un pueblo que espera la lluvia" (is not about drought, it's about a town waiting for rain).[47]

Nonetheless, it is worth unpacking the temporal reorganization behind *Cuates*'s cogent arc. González rearranges the captured footage of drought to map neatly onto the five-stage Freytag plot pyramid: exposition (normality), rising action (diminishing water), climax (drought), falling action (exodus), and resolution (rain). Drought is presented as a predictable cycle that seamlessly moves from plenitude to drought and back again. This emplotment indicates that the end of the world brought about by drought is not a singular or a definitive closure, but an ending that is infinitely reversible and repeatable, a process that is continually unfolding, a series of worlds that end and then begin again. Yet while *Cuates* celebrates the resilience of those who adapt to the rhythms of drought, it also tamps down the uncertainty of drought and makes it fit narrative expectations. By aligning rain with birth and resolution, *Cuates* delivers satisfying closure, fulfilling the promise anticipated throughout the film: the return of life. It presents drought as it used to be and as we wish it still were—a predictable temporal cycle—rather than what drought is increasingly becoming due to anthropogenic climate change: unpredictable and irregular.

Severe droughts are naturally occurring phenomena that are part and parcel of normal climate fluctuations, particularly in arid regions like northeastern Mexico. However, as with other extreme weather patterns that have intensified with climate change, regions historically prone to drought face increased risk as the planet continues to warm.[48] Current weather patterns have already been linked to climate change: the drought in northeastern Mexico captured in *Cuates* that lasted from October 2010 through September 2011 was the worst on record since the government began recording rainfall in 1941. It caused the death of an estimated 1.7 million farm animals and the loss of 2.2 million acres of crops.[49] Climate models indicate that projected warming will "significantly increase the duration and intensity of droughts" in arid regions of Mexico and lead to more extreme rainfall distribution that will decrease in the summer and increase in autumn and winter.[50] These changes will negatively impact spring-summer crops, which will require more irrigation.⁵¹ The increasing unpredictability of rainfall will imperil food security and exacerbate inequity.

The editing of drought away from this inconstancy affirms Jennifer Fay's hypothesis that cinema is an art form that manifests the human desire to control nature, in this case, the weather.⁵² González's temporal reorganization of the captured footage dissipates the uncertainty of rural futures in the era of climate change into a message of cyclical rebirth, providing the audience with reassuring certainty that drought will end and the community will continue to survive, even if that certainty is not available to the ranchers in Cuates who must continue to navigate the escalating volatility of precipitation in Coahuila. The subjunctive presentation of drought *as it should be* effectively reinforces desires for climatic stability and consequently reassures viewers that the community's adaptive migration strategies are a sufficient response, requiring no further action on the part of the state or the world. *Cuates de Australia* thus illustrates how the rural resilience film ambiguously depicts resilience by both celebrating community-driven sociocultural practices like migration and simultaneously naturalizing neoliberal systems of governance that produce subjects who accept the exposure to hazard as an individual or localized responsibility. By excluding historical markers and abstracting the ejido of Cuates de Australia into a timeless drama, González's film allows no room to consider drought as a process that is both environmentally and historically produced, connected to, and exacerbated by, neoliberalism and carbon-fueled capitalism. *Cuates* thus illustrates how subjunctive aesthetics can, at times, perform a sort of wish-fulfillment that dilutes the audience's understanding of changing climatic patterns and shores up the belief in the resilience of communities to weather drought through the reassuring narrative closure brought by rain.

RURAL ARCHETYPES

Cuates de Australia's reassuring denouement sutures the uncertainty of rural futurity in the era of climate change into archetypal immutability. These archetypes have a long history. According to Carlos Monsiváis, nineteenth-century literature framed the countryside as a bucolic space safely distanced from the moral ambiguity of the city.⁵³ The trope of rural constancy was further entrenched in the early twentieth century in the wake of the Mexican Revolution and by the rise of *indigenismo*. Artists and writers looked to the countryside as the bastion of tradition and authenticity. The site of Mexico's racial and cultural difference, the rural was at once a symbol of Mexico's past and the source of its aesthetic future, a simultaneity crystallized by films like

María Candelaria. This imagined "Mexican Eden . . . provided Mexicans with an enduring link to the countryside" and a means of escape from the mess of industrialization.⁵⁴

For bourgeois city dwellers, it was comforting to think of the countryside as an unchanging "paradise lost."⁵⁵ This impression hinged upon its perceived cultural and environmental constancy, whose "pastoral qualities [were] rendered as timeless, static in geographic, physical, and human features."⁵⁶ This reinforced the perception of the rural as a place outside of time, an unchanging font of *mexicanidad*. The rural peasantry was similarly "immobilized by such mythological representations," Monsiváis observed, out of the "desire to expunge all movement from the rural scene."⁵⁷ Juan Rulfo, he explained, famously reworked this trope in *Pedro Páramo* (1955) such that rural immobility indexed, not a static ideal, but the way in which the revolution brought such little change for rural communities.

At the turn of the twenty-first century, Mexican cinema continued to juxtapose Mexico City's claustrophobic inescapability with the *provincia* (anywhere outside of Mexico City), but now depicting the latter, as Emily Hind has argued, as "a permissive space that facilitates social freedom."⁵⁸ The provincia's emancipatory potential in the cultural imagination is linked to its greater contact with popular traditions, spatializing the allegorical reconnection with the nation and the self.⁵⁹ The counterpart to this idealized depiction of rural Mexico in cinema is its staging as a nightmarish backwater, diagnosed earlier in this chapter by Everardo González as coalescing around "hyper violent [and] bloody" stereotypes.⁶⁰ The latter cinematic tradition stretches back to the rural mobs that threaten the titular protagonist of *María Candelaria* and the brutality of the campesinos in Felipe Cazals's *Canoa: Memoria de un hecho vergonzoso* (Canoa: A Shameful Memory, 1976); it finds its contemporary analog in the political satires directed by Luis Estrada.⁶¹

If, then, the cinematic depiction of rural Mexico has vacillated between its pastoral idealization and maligned backwardness, the rural resilience film emerges from within this binary, is nourished by it, and to some extent scrambles it. The genre eschews on-screen violence, opting for a more zoomed-out tale of human versus nonhuman, as *Cuates* illustrates. Yet the next films under consideration complicate this narrative, demonstrating that the chaotic, hostile nature in question is itself manmade: the product of past human decisions. Temporally, the genre draws from the pastoral and nostalgic depiction of the rural as existing outside of history, but its focus on environmental dynamism reconfigures rural Mexico from a site visually yoked to the nation's past into a prophetic space that gestures through the subjunctive voice of the visual toward a planetary future in which environmental uncertainty is not the exception but the rule.

RURAL TIMELESSNESS AND MAGICAL RESILIENCE IN *LOS REYES DEL PUEBLO QUE NO EXISTE*

The notion that rural Mexico is frozen in time, or somehow suspended outside of it, continues to inform contemporary cinema. This is somewhat surprising given that ecodocs highlight the erosion of environmental permanence, undermining a central pillar of constructed rural constancy. Nonetheless, we see this insistence on rural Mexico's temporal excess at work in the reception of Betzabé García's documentary *Los reyes del pueblo que no existe*. Multiple reviews by Mexican outlets use phrases like "stuck in time" and "at the end of the world" to describe the town of San Marcos, Sinaloa, documented in García's 2015 film. Víctor Martínez Ranero similarly characterizes it as a "place that has been forgotten by time" in his review for *Time Out México*.[62] Yet, in a manner akin to Rulfo's critique of the revolution's failure to change life for the rural peasantry, *Los reyes del pueblo que no existe* reveals that San Marcos was actively propelled out of time to ensure the future continuity of the neighboring resort city of Mazatlán. San Marcos, then, can be understood as what Val Plumwood deems a "shadow" place: one sacrificed to better others.[63]

In 2006, construction began on the Picachos Dam to provide a thirty-year water supply for the bustling, touristed coastal city of Mazatlán, located forty-five minutes from San Marcos. Three years later, upon the dam's completion, the redirected waterway flooded six villages in its path. The eight hundred families who lived there were forcibly displaced as they watched their homes swallowed up by water. The state of Sinaloa failed to meet its legal obligation to expropriate the land in advance of the dam's construction and provide adequate compensation to the displaced for the loss of territory, instead using coercion to threaten villagers into fleeing.[64] At the time of the flooding, new construction to house the displaced was still unfinished, leaving many in limbo. The new town erected by the state lacked key infrastructure, like schools and hospitals, and its housing was smaller and more shoddily built than what it replaced. *Los reyes del pueblo que no existe* profiles three couples who chose to stay behind, rejecting their forced displacement and continuing to inhabit the flooded village. Those who did leave the flooded town also refused to accept the inadequate compensation provided by the state, collectively mobilizing into a protest movement led by Octavio Atilano Román Tirado, who promoted the *comuneros*' legal right to just compensation on his radio program *Así es mi tierra* (This is how my land is). Tirado was assassinated while broadcasting live in 2014 by forces that aimed to silence the comunero movement, a crime that has yet to be prosecuted.

García details Tirado's assassination in a companion piece to *Los reyes*, a documentary short released in 2016 in collaboration with the *New York Times* titled *Unsilenced*.⁶⁵ This short connects the dots between infrastructure megaprojects in Mexico, like Picachos Dam, and their violent imposition. It didactically presents the comuneros' forced displacement and Tirado's assassination in a way that is accessible to international viewers unfamiliar with the dispute, ending with a call for justice. By contrast, the observational *Los reyes del pueblo que no existe*, released the year prior, provides none of this context diegetically, only briefly explicating it in its closing titles. The film instead immerses viewers in the flooded landscape of San Marcos without explaining the cause behind the village's submersion until the film's conclusion, a structure that functions as a reveal but also allows the subjunctive voice of the images of life lived on water to travel via viewers' imaginations to other associations, like planetary sea-level rise.

The camera in *Los reyes del pueblo que no existe* observes its three protagonist families as they go about their daily routines—making tortillas, sweeping the church, telling jokes—against a backdrop of watery ruins. Like *Cuates*, *Los reyes* uses sensory ethnography to impart a sensorial account of the pleasures and difficulties of living in a world dictated by water, and the film resists framing its subjects as victims. What emerges is a portrait of what Lucy Bollington deems "the relational bonds [that are] woven following property's demise."⁶⁶ The families who chose to stay in San Marcos have no regular access to electricity, functional roads, health services, or the economic and emotional benefits of community. They are left vulnerable to organized crime, a lurking threat that is omnipresent but not explicitly named, materializing through the sound of gunshots heard in the near distance or through references to an ambiguous "they."

Camerawork also reinforces the notion of San Marcos as an apocalyptic site. The film opens as the camera travels in silence with a man with a lazy eye who navigates the flooded landscape in a motorboat (fig. 4.2).⁶⁷ Stark, denuded branches set against the dreary sky jut out of the water around him. As the scene progresses, the camera flips orientation to the man's point of view as he approaches San Marcos's flooded, graffitied ruins (fig. 4.3). Through mise-en-scène—the lack of sun, watery ruins reclaimed by greenery, the absence of human sound, and, most problematically, the disabled body—García establishes the apocalyptic implications of "el pueblo que no

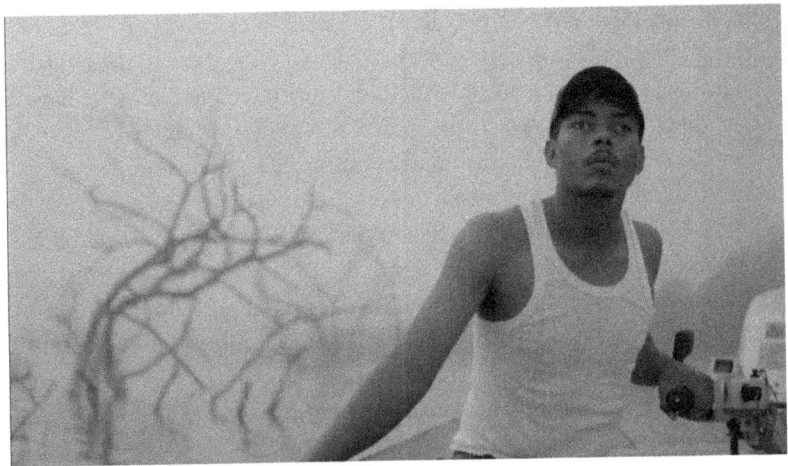

FIGURE 4.2. The introductory sequence from *Los reyes del pueblo que no existe*. 2015, Betzabé García, dir.

FIGURE 4.3. The camera flips orientation to adopt the boatman's view of the flooded landscape of San Marcos in the introductory sequence of *Los reyes del pueblo que no existe*.

existe" (the town that doesn't exist) to then subvert audience expectations as the film showcases the resilient responses of the titular kings.

As Lilia Adriana Pérez Limón indicates, *Los reyes* resists pathologizing the decision to stay in San Marcos, instead presenting small acts of self-care and daily routines of community maintenance as "ways of living better during bad times."[68] The documentary's subjects embrace life lived on water as better than forced displacement, revealed as the more dystopian of the two choices. This is encapsulated by a humorous comment made by one of the film's protagonists that García later recounted in an interview: "Allá en el nuevo pueblo les falta el agua y acá nos sobra" (Over there in the new town they lack water, and here we have too much).[69] The profiled subjects have different views: where one wants to rebuild, another feels trapped by his obligation to care for his parents. The range of expressed emotions similarly run the gamut from frustration to fear, from cynical pragmatism to hope. The film allows these varied survival strategies to coexist, indexing how resilience is divergently animated.

Of particular interest is how García teases out the temporality of San Marcos, at once presenting it as a place outside of time (outside of human history; the postapocalyptic present), a place produced by its historical moment (state-sponsored development projects that buttress the future of a coastal city at the expense of the rural), and a place that allegorically stands in for a speculative future lived on water. In what follows, I narrow in on the third of these registers to unpack how current rural realities serve as a stand-in for future possibilities of global climate change. As Falicov reminds us, the extrapolation of global themes from local realities is the dominant aesthetic preferred by the festival circuit: "a narrative that aspires to transcend borders and offer audiences 'something for everyone' across national specificities."[70]

In *Los reyes* this means using San Marcos's flooding as a microcosm that can be extrapolated through the subjunctive voice of the visual—or how images accrue connotations as they travel via viewer's imaginations to other contexts—to the postapocalyptic imaginary of climate change. The conceptual association of state-engineered flooding in Sinaloa with global sea-level rise has been present from the outset of García's project. In 2011, while she was a film student at the Centro Universitario de Estudios Cinematográficos (University Center of Cinematographic Studies) of the Universidad Nacional Autónoma de México (National Autonomous University of Mexico), García filmed a fictional short titled *Venecia, Sinaloa* (Venice, Sinaloa), her first attempt to make sense of the Picachos Dam through cinema.[71] The buoyant short depicts neighbors moving furniture out of their homes in the wake of San Marcos's catastrophic flooding. A young man takes a break from his labor to watch TV. There he sees a news report about the expanded installation of passerelles, or raised wooden walkways, in Venice, Italy, to allow citizens to navigate a city that is both sinking and frequently flooded by rising sea levels.[72] Inspired, the protagonist begins to construct similar elevated walkways in San Marcos. Another boy joins in, and the short ends with neighbors moving back into the village on the newly installed raised walkways, suggesting the community's successful adaptation to the radically transformed environment.

Venecia, Sinaloa is more utopian than either of García's subsequent films on the topic. It captures a preliminary moment of hope hinging on speculative adaptation to environmental catastrophe that eclipses the pain and violence of dispossession, which are brought into greater relief in *Los reyes* and *Unsilenced* with the benefit of hindsight. (After making the short, García moved to San Marcos for five years, ultimately directing and coproducing *Los reyes*.) It also suggests a false equivalence between the state-funded infrastructure of Venice and the DIY efforts of the people of San Marcos, who were not only abandoned by the state but actively robbed by it. Although the connection with sea-level rise is not made explicit in *Los reyes*, this early iteration that connects the dots between local manufactured crisis and global sea-level rise makes clear how García thinks of this site as exemplary of environmental catastrophe writ large, a toggling of local and global that might be a reason behind the film's success on the festival circuit.

The distribution trajectories of *Unsilenced* and *Los reyes del pueblo que no existe* reached international viewers who were likely uninitiated in Mexico's water and development conflicts. *Unsilenced* deploys traditional activist documentary techniques, like voice-over explanations and explicit audio of Tirado's murder captured during his radio broadcast, to inform *Times* viewers of the conflict's players and stakes. *Los reyes* engages a similar demographic of upper-middle-class festivalgoers, a circuit on which it was roundly successful. After receiving the best documentary prize at the Morelia International Film Festival, *Los reyes* won the 2015 Golden Eye for Best International Documentary at the Zurich Film Festival, the 2015 Reva and David Logan Grand Jury Award at the Full Frame Documentary Festival in Durham, North Carolina, and the 2015 Global Audience Award at the SXSW Film Festival in Austin, Texas. This acclaim got it picked up for distribution on transnational streaming services like iTunes and Amazon Prime—a huge achievement for a Mexican documentary. To combat accessibility issues for audiences in Mexico and elsewhere, García uploaded *Los reyes* to the free video-sharing platform Vimeo shortly after its release, where viewers can still find it today.[73]

Los reyes' success on the festival circuit is certainly due to its rich ethnographic storytelling of manufactured environmental crisis. Yet if we recall González's comment about international viewers' preference for the moving image over laborious subtitles, it isn't a stretch to suggest that its success is also indebted to its privileging of aesthetics over discourse. In sharp contrast to García's take on the same story in *Unsilenced*, *Los reyes* contains no specific political call or historicization. It is only with the expository end titles that viewers unfamiliar with Mexican current events would even be aware that this town was purposefully flooded to allow for Mazatlán's growth. Rather, through the subjunctive voice of the visual, García frames her documentary as an allegory of environmental apocalypse, an approach that makes this local tale of dispossession universalizable. In so doing, it appeals to international viewers' preexisting anxieties about climate change's implications for the future: specifically, the reality that as sea levels continue to rise, many will need to either leave their land or learn to live on water. Studies predict that even with low carbon emissions, 190 million people currently live in areas that will be underwater by the year 2100. If emissions continue at high rates—the likely scenario, given current inaction—630 million people will be affected worldwide.[74] The projected affected areas include parts of coastal Mexico, where nearly a quarter of the country's population lives. Ironically, Mazatlán, the resort town the Picachos Dam was built to sustain, is one such city that is currently experiencing shoreline erosion and is vulnerable to potentially existential future sea-level rise.[75]

García cultivates San Marcos's allegorical significance through visual tropes recurrent in stories about the end of the world. Diego Tenorio's breathtaking cinematography establishes San Marcos as a postapocalyptic space through long, wide-angle takes that dwell on the village's submerged ruins enshrouded in mist. Viewers are immersed in the ghostlike landscape, a sublime site/sight of simultaneous horror and pleasure. Because the cause behind the village's submersion is not established until the end titles, the indeterminate quality of the footage of the flooded, abandoned village invites audience speculation via the subjunctive voice of the visual, allowing the scene to be unmoored from its sociohistorical context and to take on the resonance of myth or fable. The camera accompanies villagers as they navigate what remains of San Marcos on small boats, visiting, for instance, a solitary cow trapped on a water-bound tract of land. Christian Giraud's layered sound design of nonhuman noises enhances the sense that this place is being taken back by nature, yet another motif of postapocalyptic fiction. Other moments draw from mythical tropes. A nighttime boat tour of the village captures one former resident animatedly recounting old grudges between long-gone neighbors. His barely illuminated face looms over the low-angle camera as he propels the craft forward with a long bamboo rod (fig. 4.4). This memorable sequence is a clear nod to Charon, the figure in Greek mythology who ferried souls across the river that divided the world of the living from the world of the dead.

The uncanny elements of the inundated town—a stranded cow, a submerged Christ statue, a functioning *tortillería* in the middle of nowhere—are

FIGURE 4.4. Low-angle shot in *Los reyes del pueblo que no existe* of a man who traverses the flooded town at dawn, recounting stories of neighbors who used to live there.

FIGURE 4.5. Child musicians appear as if out of nowhere, lending *Los reyes del pueblo que no existe* an air of magical realism.

heightened by the inclusion of staged elements. A prime example is the appearance of young musicians at random intervals playing banda music on instruments like the tuba (fig. 4.5). The boys do not live in San Marcos; their mysterious appearance, like much else in the documentary, goes unexplained. In interviews, García said that their inclusion provides a "toque mágico al pueblo" (magical touch to the town).[76] In other interviews, she similarly described San Marcos as a place of magical realism, a characterization that she explained she realized after seeing her film screened at the Guadalajara International Film Festival.[77] The reference to magical realism has been picked up and repeated by reviewers, from news outlets to online commenters.

The film's association with magical realism is an effective marketing strategy. This sort of a hook is a necessity for pitching a film for transnational distribution. García's discovery of these magical elements while at the Guadalajara International Film Festival is unsurprising: it is a key festival for networking and production deals.[78] Perhaps conversations with producers led her to identify this angle as a way to market her film and gain access to additional markets and funds. Yet the invocation of magical realism ties both San Marcos and the film's value to its consumable exoticism. It implicitly calls for the moral, ethical, and aesthetic recognition of San Marcos on the grounds that it is a "magical" place worth rescuing, or, alternatively, that the magic of the place is what fuels its resilience to dispossession. This is not to accuse García of disregard for her subjects; her respect for them is evident in her careful treatment of their choices and stories. But it is to signal the tension that directors working in rural spaces in Mexico must navigate as they deal with market demands and distribution networks that expect and reinforce certain narratives about what it is like to live in places made precarious by anthropogenic disasters, narratives that frame community responses to state violence and dispossession as examples of magical resilience.

Betzabé García's use of the documentary form in *Los reyes del pueblo que no existe* to speak to speculative fears of climate apocalypse forcefully demonstrates that climate crisis is not just a future concern but is already documentable in the present. *Los reyes* affords its subjects the status of seers who proffer strategies for living at the end of the world, emplotting the rural, not as belated, but as at the forefront of accelerating climate crisis. Yet, on the other hand, the subjunctive abstraction of San Marcos into an allegory of apocalyptic climate crisis disconcertingly erases the concrete political demands of the dispossessed, as well as the causes behind their dispossession: the state's active drowning of one place to ensure the continuity of another. Perhaps *Unsilenced* is García's attempt to rectify this erasure through the return to a more didactic and activist documentary style. Her use of three different narrative modes to communicate the stakes of the Picachos Dam —speculative fiction in *Venecia, Sinaloa*, sensory ethnography in *Los reyes del pueblo que no existe*, and activist exposition in *Unsilenced*—indicates that when it comes to the cinematic treatment of thorny questions of environmental justice, the demands of aesthetics and politics continue to rub up against one another in unresolved tension.

OBSERVATIONAL DOCUMENTARY AND THE INSUFFICIENCY OF THE IMAGE IN *EL REMOLINO*

The observational documentary genre, otherwise known as "direct cinema" or "cinema verité," has long been a privileged mode for representing rural Mexico, in large part because of its promise of unmediated access to the real. Documentarians using this style embed themselves for years in the community they aim to capture, an immersive experience that they then translate through the moving image to the audience. As Bill Nichols has explained, the observational mode promises immediate access to the subjects on-screen by doing away with voice-over commentary, nondiegetic music, and interviews. The assumption at the root of this mode is that the truth of a place will reveal itself if one looks long enough, privileging sight over narrative explanation. Editing reinforces this idea, Nichols writes: "The space gives every indication of having been carved from the historical world rather than fabricated as a fictional mise-en-scène."[79] Observational documentary often breaks its own rules with moments that remind viewers the film is still a construct: a direct look into the camera in the case of *Los reyes*, the utterance of the director's name in the case of *Cuates*. The meaning of the images shown on-screen invariably travels via the subjunctive voice of the visual as viewers contingently affix different connotations to them, such that an image of a town flooded by a dam in Sinaloa can come to stand in for planetary sea rise.

While observational documentary successfully captures and communicates a nuanced sense of place, it stumbles when it comes to translating environmental change. In part this is because as viewers we are not trained to see environmental dynamism. Nor, as I have argued via the narrative reorganization of *Cuates*, do filmmakers seem ready to relinquish the desire to map irregular, nonlinear environmental phenomena onto familiar narrative arcs. In *Cuates*, the result is that the linear presentation of drought on screen is naturalized, taken for granted as the unvarying truth of place, in a way that parallels the film's claim to ethnographic truth.

Because the term *climate* refers to weather patterns that unfold over an extended period, it requires abstraction beyond what is immediately evident or experienced, that is, beyond the weather or the state of the atmosphere at any given place and time. Environmental change necessitates exposition: an abstracting force that removes one from the immersive now to grasp the extended timescales of territorial dynamism. Sensory ethnographies like *Cuates* and *Los reyes*, with their focus on affective immersion, are at odds with the narrative demands of climate and environmental change, a subject that requires mediation and exposition. This abstraction can be achieved via elements such as archival footage, interviews with experts, or eyewitness testimonies, which contextualize the audience's relationship with the image but interrupt the immersive sensorial experience of place.

To illustrate how sensory ethnography can recur to eyewitness testimony and memory as narrative forces that complement and complicate sensorial immersion in place, I turn to *El Remolino*, a documentary directed by Spanish filmmaker Laura Herrero Garvín about a village in Chiapas dealing with unpredictable flooding that is the result of decades of deforestation. Like *Cuates de Australia*, *El Remolino* is named after the town in which it takes place, a village in the municipality of Catazajá, Chiapas, that lies on the banks of the Usumacinta River. The river's annual floods have grown in intensity and duration, reorganizing life into one lived on water, requiring transportation by boat, ravaging crops, and shutting down schools. The documentary follows three generations of the Benítez family who have chosen to remain in El Remolino despite adverse conditions: grandfather Edelio, his nonbinary child Pedro, daughter Esther, and granddaughter Dana (Esther's youngest child). The village has otherwise emptied out because of the river's rising waters. Of thirteen siblings, Pedro and Esther are the only ones who remain.

As with the other films discussed in this chapter, the relationship with water is the core drama of *El Remolino*; its presence delimits the plans and actions of those who inhabit this territory. Another iteration of the rural resilience film, *El Remolino* dwells on the paradoxical beauty and extremity of living in response to the ebb and flow of the flood. The subjects featured in *El Remolino* are also its storytellers; this narrative agency is formalized through the inclusion of participatory film-within-the-film footage captured by Esther and Dana. Young Dana's riddles and camcorder footage of the flood imbue the documentary with childlike wonder.

Herrero Garvín uses storytelling to dispute the naturalization of flooding in El Remolino that the image implies. In a voice-over, Dana dreamily explains, "Te voy a contar algo que mi mamá me dijo: los arboles había bastantes bastantes" (I'm going to tell you something that my mom said to me: there used to be many, many trees).[80] Her mother, Esther, elaborates that the floods are a recent phenomenon, the result of deforestation by settlers who "cut down trees to survive, not knowing what would come next." The five or six families who settled there a century ago to transform the jungle

FIGURE 4.6. *El Remolino*. 2016, directed by Laura Herrero Garvín.

into arable land catalyzed environmental transformations that would ultimately render life unsustainable in the present. Set against shots of the swelling flood, these explanations reorient how viewers see the images shown on-screen: not as a wholly "natural" or wild space, but as an environment molded by human intervention (fig. 4.6).

These narrative contextualizations explode the atemporality of the nonhuman landscape. Without the help of narrative abstraction to pull the viewer out of the immediacy of sensorial ethnography, we cannot know through image alone how quickly water arrives, how slowly it recedes, and to what extent this diverges from past patterns. Without this historicization, it is tempting to slip into the sublime of aesthetic experience detached from the quotidian particularity of being in place over time. This contextualization is provided in *El Remolino* through the lived experience and memories of the subjects on-screen, who are granted the narrative power to explicate territorial change to viewers.

In one scene grandfather Edelio stands knee-deep in water. He explains that the flooding has risen year after year by pointing to where it has reached on his body: first his ankles, then his knees, his shoulders, and now over his head. The aqueous becoming of El Remolino is too slow-moving to be caught on camera but nonetheless so rapid that it has unfolded within one generation, accounted for through Edelio's body. Whereas for one family member this transformation is a lived experience, for his granddaughter it is merely a story about the past. It is only by complementing image with memory that viewers can make sense of how this watery scene has been produced and, in turn, how it is now producing the lives of those who live with it.

The flood is presented as part of a bundle of inherited legacies of harm passed down from generation to generation, including alcoholism, sexism, and homophobia. Pedro and Esther struggle with their father's limiting beliefs, on display through Edelio's disdainful response to his children's gender nonconformity. Pedro is a queer nonbinary person, a self-described "reina de noche y de día hombre" (queen by night and man by day), who still lives with their parents. They raise chickens and tend the milpa while saving up for breast implants and a small home, hoping for a partner with whom to raise a child. Their sister Esther is a single mother who runs a small shop and wants her daughters to be well educated. Edelio, their alcoholic father, is infuriated by their nonconformity; he describes Pedro as a product of Satan and Esther as unattractive to potential romantic partners because she does masculine work.

Nonetheless, Pedro and Esther remain in El Remolino. Their decision to stay is presented as a choice. Esther explains that even if she were to emigrate to Mexico City, she would take her problems with her; by staying to face them, she suggests, they might yet be resolved. Likewise, Pedro cannot imagine leaving behind their chickens, which are their livelihood and cherished companions. Pedro describes their attachment to El Remolino as a centripetal force: life is a "remolino" that swirls them around and around, only to leave them in the same place. *El Remolino* counteracts the pervasive erasure of queer and feminist subjectivities from rural Mexico, undoing the notion that the urban is the only horizon of progressive promise or that the solution for nonconforming subjects is simply to leave.

The coalescence of gender struggles and environmental adaptation paints a complex picture of resilience in twenty-first century rural Mexico, where irregular flooding is one of many inherited legacies. In contrast with *Cuates* and *Los reyes*, *El Remolino* does not depict the rural as a refuge from capitalism or paramilitary violence, but erects a contemporary portrait of rural life in which, like the rest of Mexico, current generations wrestle with norms that are both structural and intimate: engendered by one's ancestors, instilled by one's parents. It proposes that the decision to stay and grapple with the messy host of inherited legacies is an ethical commitment. Staying means cultivating risky attachments to volatile places and people who are not necessarily nurturing but to whom we are nonetheless interlinked. Rather than disconnect from these legacies, Esther and Pablo propose that it is better to dwell in them, remember fraught histories, and slowly unknot them. This is what Donna Haraway calls *ongoingness*: "nurturing, or inventing, or discovering, or somehow cobbling together ways of living and dying well with each other in the tissues of an earth whose very habitability is threatened."[81] In this sense, *El Remolino*, like *Cuates* and even *Los reyes*, argues that the future of rural Mexico, while imperiled, is not foreclosed. Attachments to likely and unlikely kin (children and chicken) bind Pablo and Esther to El Remolino, a manufactured environment that requires adaptive ways of dwelling in response to the contingency of water: modes of cultivation, caretaking, and maintenance that are carried out in recognition of their potential unraveling.

UNCERTAIN FUTURES

The destabilization of life in rural Mexico set in motion by NAFTA, which disproportionately affected small-scale farming by flooding the market with cheap corn and bean imports, has in the twenty-first century entered a new stage as anthropogenic climate change alters the logic of the environment itself.[82] As these three films indicate, the scarcity or excess of water is one lens through which to bring disruptive environmental dynamism into focus. Rural territorial transformations are divergent in origin—deforestation by landholding ancestors, a dam built to sustain a resort town, global warming—but all point to the insatiable demands of capital. The localized apocalyptic effects of environmental crisis pose yet another challenge for communities already grappling with violence, poverty, impunity, and other iterations of state neglect.

The rural resilience film breaks with the rural's stereotypical positioning in Mexican cinema as belated or timeless. Far from it, the rural portends the catastrophic future of accelerating climate change, already evincing its effects as well as inspiring potential strategies for survival. We see this temporal elasticity at work in *Los reyes del pueblo que no existe*, which subjunctively deploys San Marcos's inundated present as allegory for a planetary future lived on water. Even as environmental dynamism makes it difficult to keep thinking of the rural as a time capsule of authenticity, it continues to hold this allure, as *Cuates de Australia*'s heavy dose of nostalgia indicates. The genre of rural resilience is thus a contested arena where competing feelings about the import of the emptying out of the rural are waged.

To recapitulate, the rural resilience film is characterized by a set of shared concerns. It trains its focus on an isolated, reduced set of individuals who have chosen to stay in a location defined by hostile environmental conditions. In this sense, it can be understood as the other side of the coin to migration narratives. A much more prolific genre, the cinema of migration focuses on individuals compelled to leave their homes in search of something better. But while these films flesh out the various motivations that propel people to leave, only infrequently do they account for those who remain behind, whether by choice or out of necessity. The rural resilience subgenre takes up this task, foregrounding the resolve to remain in place as a decision that is as world building as the decision to emigrate.

In *Cuates de Australia*, *Los reyes del pueblo que no existe*, and *El Remolino*, remaining in place is framed as an active choice, albeit one that some subjects resent. This mirrors the tendency of migration narratives to present migration as a difficult choice necessitated by unsustainable circumstances. So too does the rural resilience subgenre depict the determination to remain in abandoned or hostile territories as an agential wager on the future of territory, even when there is no indication that environmental conditions will improve. Indeed, in most cases, the opposite is true: precipitation in *Cuates de Australia* will only grow more irregular, and flooding in *El Remolino* will continue to endanger crop cultivation. The profiled subjects understand and accept these stakes, illustrating the framework of "futurity without optimism" put forth by Allyse Knox-Russell.[83] The willingness to cultivate a future amid precarious conditions is a bet on an existence that might not look like the "good life" prescribed by capital, but one that can be returned to when rain permits, one that can be adapted to by living on water, or one revolving around affective attachments to unlikely kin.

Another characteristic of the rural resilience film as outlined here is the obviation of the stereotypical focus on rampant violence that often accompanies contemporary films about rural Mexico. By removing bare violence from the frame (or only alluding to it in passing), filmmakers resist reducing rural dynamics to the tired and misleading explanation of the drug war.[84] They also studiously avoid sensationalism, instead turning to sensory ethnography to communicate the nuanced feel of inhabiting contexts of rural ruin. In this regard, foregrounding environmental crisis as inextricable from other socioeconomic factors that have rendered rural living increasingly precarious is a key intervention of the genre, in that it wrests focus away from organized crime as the sole cause behind the upending of the rural.

This brings us to the knotty question of resilience. On the one hand, these films point to rural survivors of environmental crisis as modeling strategies for living and even flourishing amid adverse environmental conditions, lessons in adaptation that are needed in the era of accelerating planetary crisis. The resilience on display is collectively engendered and imagined from the bottom, troubling the notion that rural subjects are passive victims in need of saving by top-down liberal governance. The cyclical migration practiced by ranchers in *Cuates de Australia*, for instance, models a solution to water insecurity that is community driven, doesn't require abandoning territory, and is managed without state intervention.

Yet the celebration of community adaptation to climate crisis without state support points to a conundrum of such films and the concept of resilience in general. Resilience has been robustly critiqued for how it has been strategically mobilized by neoliberal power to deflect state responsibility for the prevention and management of economic, environmental, and social crisis. Those who have been harmed by the state, like the comuneros forcefully displaced from San Marcos in Sinaloa, are tasked with fending for themselves. Championing individual resilience runs the risk of mirroring the neoliberal maxim that citizens should "accept the necessity of living a life of permanent exposure to endemic dangers" and come to terms with the world as it is, rather than putting the onus back on the state to assume responsibility for rural livability.[85]

The discourse of resilience in the rural resilience film reflects the divergent and ambivalent valences of resilience. It dramatizes the way in which a concept like resilience can be deployed to celebrate quotidian acts of maintenance amid situations of environmental damage, but also the way in which stories of individual resilience can be co-opted to abdicate the collective and state responsibility to respond to climate crisis. Like other forms of contemporary Mexican cinema, the genre of rural resilience runs the risk of romanticizing deprivation and slipping into the sublime in its aesthetic treatment of ruined, dystopian, or increasingly hostile rural environments. We see this at work in both *Cuates de Australia* and *Los reyes del pueblo que no existe*, where the dynamics of festival circulation incentivize aesthetic immersion and tropes of magical resilience over narrative exposition. In such films, the subjunctive voice of the visual flourishes; the lack of expository information assigned to images of hostile environments means that they are unfixed, open to interpretation by the audience, who fills in the gaps through emotions, opinions, fears, and imagination. The results of this subjunctive invitation are ambiguous: it can cultivate generative interpretative leaps that connect the state erasure of one place to ensure the future of another with planetary processes of climate change, an analog that underscores the racialized and classed dynamics of dispossession, but it also runs the risk of eroding the specificity of the documented scene, allowing audiences to see what they wish to see rather than what is really at stake. This is why thinking about the dynamics of distribution and reception is crucial to understanding how ecocinema accrues meaning. While the rural resilience film's sensorial immersion in place provides a sense of proximity that can compel renewed ethical engagement between spectator and subject, this illusion of proximity can also obscure the film's mediation of rural crisis as a commodity to be sold to urban audiences. What does it mean that most of the people who watch the films of rural resilience are not rural themselves? It is to this question that I turn in the next chapter.

CHAPTER 5

Greening Mexican Cinema

In the summer of 2021, a web series debuted on YouTube titled *El tema* (The Issue), organized around the tagline "El tema es el clima" (The issue is the climate). Created by movie star and producer Gael García Bernal in collaboration with Pablo Montaño Beckmann and produced by La Corriente del Golfo, a production house cofounded by García Bernal and Diego Luna, the series makes the case that climate change is *the issue* that Mexico faces in the twenty-first century. Each of the six ten-minute episodes is oriented around a specific topic (another possible translation for *tema*) and place: from water in Chihuahua to carbon in Coahuila. In each installment, García Bernal and his cohost, Mixe public intellectual Yásnaya Elena Aguilar Gil, guide viewers through the issues, speaking on the ground with activists, scientists, and engaged citizens. The series mobilizes tropes characteristic of the contemporary ecodocumentary, like García Bernal's star power and aerial shots of denuded forests, to make the case that environmental crisis has not yet received enough mainstream attention in Mexico, while at the same time highlighting the relentless efforts of those already advocating for better presents and futures for their communities.

Most interesting for the purposes of this chapter is *El tema*'s format. Two decisions in particular stand out. First, even though its content collectively clocks in around eighty minutes, *El tema* was not released as a feature-length film but as a series of bite-sized episodes. And, second, *El tema* debuted directly on YouTube rather than through distribution channels commonplace to Mexican environmental cinema, such as specialized festivals, art house movie theaters, or cable TV. Often, YouTube or Vimeo is the final destination for such films, where they are uploaded either licitly or illicitly in order to reach wider publics after traversing the festival circuit.

Yet, thanks to backing from La Corriente del Golfo, *El tema* shortcut this route, landing directly online in weekly installments between April and May of 2021, available to watch for free. This distribution strategy maximized the series' reach, enabling viewers to easily share episodes with their networks on social media. As of this writing in August 2023, the first episode had garnered 113,000 views, a sizeable audience for the genre of Mexican educational documentary.[1]

La Corriente del Golfo's decision to produce *El tema* as a web series and to distribute it online responded to ongoing COVID-19 pandemic-related health concerns associated with enclosed exhibition venues in the spring of 2021 but also to the realities of film consumption in Mexico. Although Mexico has the world's fourth largest film exhibition infrastructure in terms of the sheer number of screens, 90 percent of movie theaters in Mexico are owned by just two companies, Cinépolis and Cinemex.[2] As a result of the deregulation of film distribution codified into law by NAFTA, these companies are beholden to contracts that require that nearly 90 percent of screened films are Hollywood productions. This arrangement allows Hollywood to leverage captive global markets in places like Mexico to recoup costs for domestic flops like *Sonic the Hedgehog*, the most widely seen movie in theaters in Mexico in 2020.[3] Moreover, Cinépolis and Cinemex theaters are consolidated in upper-middle-class urban neighborhoods. The elevated cost of a movie ticket, approximately 70 percent of a day's earnings on minimum wage, accounts for the saturated spatialization of exhibition infrastructure. All of this has made moviegoing in twenty-first century Mexico highly regionalized and classed, or what Juan Llamas-Rodriguez characterizes as a globalized luxury experience.[4] By contrast, most Mexicans consume Mexican cinema on television, on pirated DVDs, as BitTorrent downloads, or on YouTube.

In this context, *El tema*'s free online distribution was a choice that aligned with its activist messaging of environmental justice as an intersectional issue of equity and access. Even so, access to high-speed internet capable of streaming content like *El tema* is far from democratic in Mexico, mapping onto similar spatialized socioeconomic inequities. Considering environmental media like *El tema* in tandem with the infrastructures of film distribution helps bring into focus key questions about ecocinema in the twenty-first century: Whom is it for? And what are the environmental implications of the increasingly atomized consumption of cinema?

This chapter follows the line of argumentation that I began to flesh out in Chapter 4, where I addressed the uneven circulation of documentaries about rural resilience to largely urban festival publics. Here I continue to think about reception dynamics to make the case that the infrastructure of cultural consumption is a salient object of study for scholars of the environmental humanities. As such, this chapter entails a methodological shift. Whereas previous chapters explored various formal properties of subjunctive aesthetics, this chapter adopts a sociological approach to culture in the era of climate change, so that we might consider subjunctive practices in postcarbon filmmaking and film exhibition that imagine the infrastructures of cinema otherwise.

If one of cinema's inherent promises is its ability to give ephemeral form to a shared space, or commons, around which different subjects can come into contact and possible relation, it is infrastructure that makes this coming together possible—from the structures of funding that finance film productions to the physical space of the movie theater, from the electronic bandwidth necessary to stream films to the distributors and telecommunications systems that facilitate their circulation. Infrastructure wields organizational force. As the last chapter illustrated through the example of the Picachos Dam in Sinaloa profiled in Betzabé García's sensorial ethnography *Los reyes del pueblo que no existe*, infrastructure is capable of ending worlds (the rural towns flooded by the dam) and prolonging others (the tourist town of Mazatlán, whose future the dam ensured). While the infrastructures of culture do not leverage the same socioecological force as extractive megaprojects, they nonetheless subtend the stories that are told and the ways that they are consumed. "Settler colonial systems of value," write Winona LaDuke and Deborah Cowen, "are literally, physically, enabled by infrastructure."[5] LaDuke and Cowen point out that by the same token, infrastructure is a potent site for building alternative models of sociality that can enable the flourishing of the commons rather than its fragmentation.

The infrastructures of cultural production therefore present an opportunity for subjunctive reimagining. How might the lessons learned from the stories told in the era of climate change outlined thus far in this book, ranging from robust critiques of extractivism to affirmative practices of land defense, be applied in service of reimagining the infrastructures of cultural production? What might cultural infrastructures built to sustain a world after oil look like? While these questions can be approached from other forms, like literature or visual art, Mexican cinema provides a particularly fruitful starting point. This is because the film industry, more so than any other form of cultural production, requires extensive infrastructure at every stage, from execution to exhibition. Moreover, the film industry is more energy intensive than any other cultural form.[6] In response, subjunctive experiments in postcarbon infrastructure aim to reimagine and upend the material and social ties between cinema and oil; they reinsert doubt and uncertainty around questions of how cinema should be made and consumed. In so doing, they underscore the medium's contingent relationship to fossil-fueled modernity and its potential to be reconfigured in a time of accelerating emissions. Experiments in postcarbon cinema thus frame cinema as both a modality of cultural expression and critique as well as a space of active experimentation in storytelling practices that can be aligned with the precepts of degrowth.

The methodological turn to infrastructure offers an opportunity to expand how we study Latin American ecocinema beyond questions of representation to consider questions of praxis and industry, that is, the material realities of film production, distribution, and exhibition. Within the field of Latin American film studies, these questions have been siloed. The growing body of scholarship that attends to environmental cinema and media in Latin America has focused primarily on how cinema visualizes and narrates environmental crisis.[7] Such a focus sheds light on how cinema operates in tandem with environmental activism or makes postextractivist ontologies palpable. But when we think about the relationship between the environment and cinema, it isn't just content that matters. Cinema is an industry that is both "culturally *and* materially embedded."[8] Illustrative of the material turn in cinema and media studies is Jussi Parikka's portmanteau word "medianatures," which underscores that these two terms are inextricable: media does not just shed light on nature, it is made possible by nature and leaves long-lasting material legacies in its wake.[9]

While there has yet to be a rigorous accounting of the materiality of media in Latin America, the field has long attended to the structures of coloniality and imbalanced global dynamics embedded in cinema. Scholars have demonstrated the generative potential of studying cinema as an industry, which, like other industries, operates within national and transnational flows of culture and capital. On a parallel track, Marxist interpretations have shed light on cinema's social reach, as well as its dependence on capital and the state. If a recurrent nexus for Marxist scholars has been to interrogate the relationship between cinema and socioeconomic conditions, ecocritical engagements with this tradition can extend such questions to consider the role of extractivism in these circuits. One pioneering example of such efforts is *The Magical State*, Venezuelan anthropologist Fernando Coronil's assessment of how oil transformed the Venezuelan state. Sean Nesselrode Moncada and Santiago Acosta have built on this work to consider how the midcentury influx of oil impacted Venezuelan cultural production.[10] Other scholars who have discussed the materiality of Latin American cinema include Martina Broner. In her study of the Colombian art house film *El abrazo de la serpiente* (Embrace of the Serpent, 2015, directed by Ciro Guerra), Broner argues that the use of salvaged 35mm film stock, "a format in extinction," echoes "the forest's fragility."[11] In a similar vein, Sebastián Figueroa has linked the global emergence of cinema with nitrate (a material used to make early film stock) extracted from the Atacama Desert.[12] Following such efforts, I propose that thinking materially about Mexican cinema—in both the ecocritical and the Marxist senses of the word—reveals its organizational and spatial force as well as subjunctive opportunities for its reimagining.

The fact that the minerals and other raw materials that have made cinema possible have been largely sourced from the Global South makes the analytic of materiality particularly relevant for scholars of Latin American cinema. Here I follow Cajetan Iheka's observations about Africa, where the promise of media—as a mechanism of cultural expression and critique, as a connector of peoples and cultures, as a sign of modernity—coexists alongside the devastating realities of its ecological costs and consequences. The tension between the affordances and the shortcomings of media, Iheka posits, is central to rethinking "ethical living and media production in a time of finite resources."[13] The same is true in Latin America, where cinema has diversified and proliferated as the technologies of filmmaking and film watching have become more portable and affordable, opening the field to historically marginalized filmmakers and allowing viewers to engage with content, like *El tema*, that has been sidelined by the mainstream media without needing to traverse formal distribution routes. At the same time, the proliferation of devices devised around planned obsolescence—smartphones, laptops, flat-screen televisions—on which to stream content like *El tema* is another key thread in the story of contemporary Latin American cinema.

In what follows, I undertake an abbreviated materialist sketch of cinema in Mexico. First, I touch on how the infrastructures of film have intersected with the infrastructures of oil. Next I turn to access, the biggest problem facing Mexican cinema today. I ask what is gained and what is lost when cinema becomes increasingly atomized. Then, I look at an example that models how the industry is beginning to imagine postcarbon cinema exhibition and makes corporeal the connection between cultural consumption and energy: Cine Móvil ToTo. I conclude by zooming out to consider the global complexities of greening cinema through a brief discussion of runaway productions and a few examples of filmmaking practices that reduce emissions while centering justice and equity.

CINEMA AND OIL

The fourth episode of *El tema*, "Energía," focuses on the state of Tabasco, home to numerous oil fields and the controversial Dos Bocas megarefinery managed by state-owned oil company Pemex (Petróleos Mexicanos), which was under construction at the time of the episode's filming. The episode narrates how Pemex has transformed the region into a sacrifice zone: endangering mangrove forests, polluting waterways, and jeopardizing fishing. A title card explains that climate change compounds this vulnerability: sea-level rise threatens to flood nearly a quarter of the state over the next thirty years, including the site of the Dos Bocas refinery. The state's decision to build a new oil refinery in an area that is projected to be underwater in just a few decades underscores the impasse between short-term investment in fossil fuel infrastructure and the territory's long-term future, which is jeopardized by the very emissions that this sort of infrastructure makes possible (figs. 5.1 and 5.2). While infrastructure is often thought about as an investment in the future, the planned obsolescence of the Dos Bocas refinery demonstrates an opposing approach to infrastructure organized around the myopic demands of capital and its short-term profit horizons.

The *El tema* episode highlights local efforts to combat regional foreclosure, including the restoration of mangrove swamps (a key regulator of flooding) and advocacy for alternative, postextractivist forms of energy production. Host Yásnaya Elena Aguilar Gil argues that a post-oil Mexico could be enacted from the ground up, explaining, "esas energías limpias

FIGURE 5.1. Against a backdrop of Gulf mangroves juxtaposed with the under-construction Dos Bocas refinery, the postscript for episode 4 of *El tema*, "Energía," reads: "For more than fifteen years the production of oil in Mexico has been in decline. Oil will run out." "Energía," *El tema*, 2021.

pueden estar gestionadas de otra manera que no sea el capital o el estado" (those clean energy forms can be managed other ways, not just by capital or the state), through small-scale, communal efforts that "[apuestan] por la vida" (bet on life), rather than death.[14] The episode calls on the audience to join in the collective task of post-oil imagining and conjure up forms of infrastructure—assemblages that bring peoples and territory into relation—to sustain future life in coastal Tabasco.

El tema illustrates how cinema can shed light on the socioenvironmental toll of extractivism and amplify activist calls to subjunctively imagine energy sovereignty beyond fossil fuels. But what if we reverse the terms of this equation? Rather than consider how cinema illuminates energy, how does studying cinema through the lens of energy change our understanding of cinema as an industry and an artistic practice? How might we extend Aguilar Gil's call for postcarbon energy production to reimagine the distribution and production of cinema in the twenty-first century?

As I have argued elsewhere, there is much to be gained by considering the structural ties between Mexican cinema and state oil revenues.[15] While perhaps to a less comprehensive extent than countries like Venezuela or Nigeria, oil has buoyed cultural production in Mexico. Ever since the industry was nationalized in 1938—a radical gesture that wrested control of subsoil resources from US oil companies and gave it back to the people of Mexico—hydrocarbons have bolstered national development. Up until recently, Pemex has essentially functioned "as the government's cash cow," steadily providing close to 40 percent of the federal government's operating revenue.[16] This financial reliance on subsoil energy reserves means that all state-funded enterprises, including cultural industries, are enmeshed in the waxing and waning of global energy prices and supply. Put differently, oil is an elemental

FIGURE 5.2. The episode's postscript continues: "Pemex is incapable of bringing back the Mexican 'miracle.' But it can continue to accelerate climate crisis." "Energía," *El tema*, 2021.

media that has mediated the shape and scope of Mexican cultural industries. Understood in this light, Mexican cinema comes into view as petrocinema.

Like other countries in Latin America and around the world, historically the Mexican film industry has been sustained by state funding. However, since the 1980s, the private sector has gained a strong foothold. Today, only about half of all Mexican productions receive some form of state support. The neoliberal shift of the film industry away from public funding was triggered by the 1982 debt crisis and cemented by the subsequent "lost decade" of economic depression. The debt crisis was catalyzed by plummeting global oil prices and demand. Just a few years earlier, the discovery of vast oil fields had raised hopes that Mexico would become "the new Iran," leading José López Portillo's administration to invest heavily in costly extractive technologies and rapidly expand oil production.[17] By 1982, oil constituted a massive 72.5 percent of Mexico's total exports, and the administration heavily relied on oil futures to leverage state borrowing. When the global price of oil dropped later that year, the economy went into a nosedive, and with it, so did state funding for cultural industries.

In the wake of the oil crisis, the Mexican film industry was radically transformed by austerity measures. Up until that point, as Ignacio Sánchez Prado writes in his study of this neoliberal sea change, cinema in Mexico had been characterized by "a sprawling State-centered infrastructure, sustained by a network of government-owned theaters, strict controls on ticket prices, mandated percentages of screen time for Mexican films, and vast subsidies for film production."[18] This infrastructure reinforced a philosophy of cinema as a shared social and cultural experience that cut across demographic and geographic differences, a philosophy that transmuted as the industry underwent privatization. The deregulation of film admission prices, domestic distribution quotas, and exhibition management was later codified into law by NAFTA. As a result, the cost of attending a movie rose and became prohibitively expensive for large swathes of the population. Sánchez Prado has shown how these changes radically altered the content of Mexican cinema, which shifted to appeal to urban upper-middle-class consumers who could afford to attend the movies.[19]

STRATIFICATION AND ACCESS

One of the weightiest consequences of the neoliberal transformation of cinema in the wake of the oil crisis has been the stratification of access. According to the *Anuario estadístico de cine mexicano 2020* (2020 Statistical Yearbook of Mexican Cinema) published by the Instituto Mexicano de

Cinematografía (IMCINE, Mexican Film Institute), 88 percent of municipalities lack a movie theater, meaning that forty million Mexicans are unable to easily access an exhibition space. This plays out across a rural-urban divide, with nearly a quarter of the nation's exhibition infrastructure located in Mexico City. Even within Mexico City, movie theaters tend to be concentrated in wealthier, centric areas, reflecting the tacit reframing of filmgoing as an activity for the middle and upper classes. Ana Rosas Mantecón describes cinema's neoliberal transformation as part and parcel of widespread "processes of spatial, social, economic, and cultural segregation" in twenty-first century Mexico.[20] In a recent assessment of this situation, the research commission of the Comunidad de Exhibición Cinematográfica (CEDECINE, Community of Cinematographic Exhibition) described access as "the most serious problem facing Mexican cinema," noting that most domestic film releases debut only in specialized theaters in upscale areas of Mexico City.[21]

This context similarly shapes Mexican ecocinema, broadly defined as cinema with environmentalist content, a genre that has exploded since 2010. One indicator of stratification in Mexican ecocinema is the high level of rural representation in films that are made by, and circulate among, urban publics. As discussed in Chapter 4, feature-length films like *Cuates de Australia*, *Los hombres del pueblo que no existe*, and *El Remolino* document developmentalism's harmful socioenvironmental impacts on rural towns and highlight the resilience of those who choose or are forced to stay in situations of environmental crisis. These well-crafted, affecting films have been successful on both the domestic and the international festival circuits, as well as at specialized environmental film festivals.

Yet despite the powerful representation performed by these films to raise awareness of the struggles of those who have been dispossessed or unimagined by the state and global capital, the fact remains that rural peoples are largely unable to access exhibition spaces to see such films. Thus, while ecodocumentaries like these might be lauded for their ability to capture and communicate environmental crisis, thinking at the level of exhibition underscores how regimens of seeing are made possible for some and not for others, a dynamic *El tema*'s distribution strategy seeks to redress. This strategy comes into heightened relief when we consider that the infrastructures of extractive ventures that have caused environmental crisis (oil pipelines, hydroelectric dams, and so on) have been imposed on the same areas where the infrastructures of film exhibition have been eroded. Cinema therefore is not just a modality of critique of the inequities that are structural to extractivism; it also retraces and reifies capital's spatialized logic of unequal exchange through its flows and infrastructures of production and consumption.

The steep decline of exhibition spaces in rural Mexico and the increased availability of handheld devices inevitably leads to the question of whether streaming has solved, or will solve, the problem of access. Indeed, the expansion of information and communications technology infrastructures throughout Mexico has made the contemporary cinematic experience increasingly digital, as we see with *El tema* on YouTube, or with the state-run streaming platform FilminLatino, where many ecodocumentaries and other domestic art house films can be viewed for an affordable fee. And yet telecommunications infrastructures are characterized by imbalances similar to those of exhibition infrastructures. According to the Instituto Nacional de Estadística y Geografía (INEGI, National Institute of Statistics and Geography), internet access in Mexico continues to operate along an urban-rural divide. Only 19 percent of homes in rural Mexico report internet connections, although 50 percent of rural inhabitants report internet access (compared to 78 percent of urban inhabitants), which likely correlates to the increased availability of affordable or repurposed smartphones.[22] Complicating the use of streaming are poor connection speeds, which are particularly problematic in southern Mexico, like in Tabasco, the state that *El tema* profiles. The slow development of robust telecommunications infrastructure has been attributed to challenging topography, low population density, and the lack of incentives.[23] What is worth underscoring here is that content like *El tema*, even with its democratic distribution strategy on YouTube, continues to reach primarily publics that are other to the subjects who are depicted on screen.

So what does the decline of movie theaters and the upswing in handheld devices mean for the film watching experience and its ecological footprint? In her panoramic assessment of exhibition in Mexico, Rosas Mantecón is quick to point out that cinema has not always been a collective experience. Cinematic precursors like the magic lantern and the Kinetoscope were designed for individualized bourgeois viewership.[24] And yet, the contemporary global shift in film distribution and exhibition away from the theater and toward streaming is not only a return to a more atomized, stratified viewing experience; it is also incomparably more carbon intensive. In their discussion of the high level of emissions associated with high-resolution streaming, Laura U. Marks and Radek Przedpelski explain that, although streaming is often described as decolonizing cinema in its facilitation of access to historically marginalized cinemas, in fact the universalization of streaming can also be understood as a form of "bandwidth imperialism," because of its prescriptive imposition of a resource-intensive model of cultural production and consumption that is ultimately unsustainable at a global scale.[25] High-resolution streaming is premised on the fantasy of limitless capacity and the denial of data's material costs. Under current fossil-fuel energy regimes, the expansion of high-resolution streaming is unfeasible without significant ecological costs: in 2019, streaming accounted for 1 percent of global greenhouse gas emissions; researchers project that by 2030, it will be responsible for 7 percent of global emissions.[26] Therefore, rather than uphold high-resolution streaming as the gold standard that should be adopted by all, Marks and Przedpelski propose an inverse path, charted by Global South filmmakers who embrace low-resolution, small-file aesthetics.[27] While this is perhaps driven by necessity, it is a model that could be purposefully adopted by filmmakers interested in aligning their work aesthetically and ethically with the precepts of degrowth, or the "redistributive downscaling of production and consumption . . . as a means to achieve environmental sustainability, social justice, and well-being,"[28] a philosophy that challenges market-based approaches that hinge on ever-escalating energy and material costs.

The global trend toward streaming brings other contradictions into view. On the one hand, the expansion of affordable portable electronic devices has made filmmaking more democratic.[29] The explosion of Latin American features and shorts dedicated to environmental issues can be linked to increased access to inexpensive filmmaking equipment, editing software, and other necessary resources, all of which has made it easier for historically marginalized filmmakers to bypass traditional funding and production routes that have favored well-educated, urban white men. On the other hand, the expansion of low-cost electronic equipment with short life spans is correlated to the exponential growth of electronic waste. The transnational dynamics of e-waste disposal reify entrenched colonial logics, tracing a trajectory that inversely mirrors the extraction of minerals from the Global South, which is a condition of possibility for industries like Hollywood. Mexico, for instance, is the primary global exporter of flat-screen TVs and the top importer of used or defunct electronics like smartphones from the United States.[30] The informal disposal of e-waste in Mexico exposes industry laborers and their communities to harmful heavy metals, resulting in a slew of ills for human health and the environment. To holistically consider the global rise of streaming and smartphones entails acknowledging where these devices originate and end up: where they are shipped, sorted, resold, repurposed, deboned, smelted, and dumped. It is also to recognize the environmental effects of media as not only affectively moving viewers

FIGURE 5.3. Cine Móvil ToTo set up for a screening in Tabasco in 2021. Courtesy of Cine Móvil ToTo.

to care about the environment through pathos but as materially touching waste workers, informal recyclers, and their children through dangerous environmental by-products.

In Mexico, both cinema and infrastructure broadly understood have been conceptualized as tangible signs of modernity and development. As my abbreviated account has shown, these two forms of production have frequently operated in tandem, with oil revenue bolstering cultural industries and cultural industries reinforcing the values of petroculture as coequal to prosperity. The shrinking of exhibition infrastructure that has accompanied the waning of oil reserves underscores the contingent correlation of these two forms and the perils of national reliance on oil to sustain cultural production and other social programs. Countries throughout Latin America have used extractivism's ability to fund social programming as a shield to deflect critiques of its socioenvironmental toll.[31] In Mexico, the acceleration of state investment in drilling and refining infrastructures and exploratory fracking in the hope of reviving oil as a central source of revenue at the same time that the state withdraws support for cultural infrastructures signals the perpetuation of this model. Untangling the knot of cinema and oil—to take one step further *El tema*'s call to viewers to confront the inevitable exhaustion of oil and to come together around reimagining small-scale, community-led "alternative forms of clean energy"—requires that we imagine new forms of cultural and energy infrastructures that might bring a world beyond fossil fuels into being. I now turn to one such experiment, Cine Móvil ToTo.

FIGURE 5.4. Attendees at a Cine Móvil ToTo screening in Tabasco in 2021 pedal the ToTo bikes. Courtesy of Cine Móvil ToTo.

CINE MÓVIL TOTO AND POSTCARBON EXHIBITION

In response to the neoliberal consolidation of film exhibition and the consequent crisis of access in Mexico, communities, nonprofits, and filmmakers have patched together alternative exhibition infrastructures.[32] In her analysis of activist filmmakers in Mexico, Freya Schiwy explains that activists invested in reaching historically marginalized film audiences distribute their films in ways that "mix older with newer production and delivery platforms... screened in community centers; broadcast on terrestrial television; or streamed on YouTube, Vimeo, Indymedia, or other internet sites."[33] Cine Móvil ToTo is an analogous initiative that borrows tactics from established models of ambulant theater and updates them to subjunctively reimagine exhibition for the era of accelerated climate crisis.

Cine Móvil ToTo is a nonprofit that seeks to redress the lack of film distribution in rural Mexico and simultaneously decarbonize exhibition. It does so by bringing bike-powered, open-air film screenings of Mexican movies to rural communities that lack movie theaters or streaming capabilities. Founded by Roberto Serrano and Diego Torres in 2013, Cine Móvil ToTo really took off in 2019, when corporate sponsorships enabled them to scale up their efforts from twenty ambulant screenings per year to more than two hundred.[34] The organization's mission to bring Mexican cinema to rural populations was based on Torres's previous experience with El Cine en tu Comunidad (Cinema in Your Community), a now-defunct IMCINE (i.e., state-funded) ambulant exhibition program that similarly aimed to expand access to Mexican cinema. Serrano and Torres's twist on this model was to power it with renewable energy.[35]

Serrano and Torres explain that the bike-powered model of Cine Móvil ToTo was inspired by a documentary, Sergio Morkin's *Los Ginger Ninjas, rodando México* (The Ginger Ninjas Ride Mexico, 2012), about a rock band that toured Mexico and California by bike. The Ginger Ninjas also used bikes to power their concerts, involving the audience in pedaling to charge the band's generators and in turn power their music equipment. To adapt this model for film exhibition, Serrano and Torres hired the band's sound engineer, Dante Espinosa, to create their bike-converter prototype, which they crowdfunded in 2013.

The basic operations of Cine Móvil ToTo are as follows. A van-based team conducts regional tours with numerous prescheduled stops for screenings in towns and villages. When they arrive, in conjunction with local governments and community organizations, the ToTo team determines the best space for the screening, typically a public plaza or open-air sports court with enough room to project a film for a seated audience. The outdoor nature of these venues has been a plus as Cine Móvil ToTo has continued to operate during the COVID-19 pandemic, in contrast with traditional exhibition venues that have closed temporarily or permanently. The host community provides seating for the show in the form of benches or plastic chairs. At the front of the audience, ToTo sets up four stationary bicycles. In advance of and during the show, community members, often children, pedal the bikes to power the exhibition equipment (figs. 5.3 and 5.4).

The ToTo team drums up viewers through prior publicity, as well as on the day of the event through announcements over a loudspeaker attached to the van, affectionately called the ToToneta. Armed with a catalog sponsored by IMCINE of a dozen domestic films, the team gives a brief synopsis of the available options to the public. The public then votes on which film to screen. Often, family-friendly films prevail, like Alberto Rodríguez's animated film *La leyenda de la Llorona* (The Legend of La Llorona, 2011), since Cine Móvil ToTo tends to draw families and young people. Other films featured in the 2021 offerings included *Tío Yim* (Uncle Yim, 2019), a documentary by Zapotec filmmaker Luna Marán; *Alamar* (2009), Pedro González Rubio's fictionalized ecodoc about a father who teaches his son to fish on Mexico's largest coral reef; and *El sueño del Mara'akame* (Mara'akame's Dream, 2016), director Federico Cecchetti's feature film that follows a Wixaritari teen torn between his dreams of becoming a rock star and his father's plans for him become a Mara'akame, or Huichol shaman. The fact that most of the movie listings on offer in 2021 featured Indigenous subjects and languages reflects ToTo's interest in selecting films that might be of interest to rural and

Indigenous audiences. These are also films that are unlikely to reach rural villages in the form of pirated DVDs, which favor bigger-budget commercial films from Hollywood and Mexican studios.

Cine Móvil ToTo is the latest iteration in a long tradition of Latin American *cines móviles* (mobile cinemas) first launched in Cuba in the 1960s. Tamara Falicov explains that the model was based on Russian agitprop trains that toured the country after the 1917 October Revolution, bringing artists and actors to rural Russians to disseminate Communist policies and propaganda. In Cuba, cines móviles were similarly conceived as a crucial part of Fidel Castro's nation-building project. Films and projectors were transported to remote locations by "trucks, mules, and even fishing boats" to signal the Cuban Revolution's commitment to bringing modernity to rural regions and as "a means of communication with a historically disenfranchised population."[36] As Falicov notes, the program was a success, tripling the number of film viewers in Cuba within a decade.

Cine Móvil ToTo exemplifies how the ambulant model has been refashioned in the twenty-first century, when access to cinema endures as a central issue but exhibition is no longer considered within the scope of the neoliberal state. If mobile cinemas have historically functioned as a means of spreading propaganda and consolidating sentiments of national belonging, Cine Móvil ToTo carries on in this tradition. The state continues to underpin the initiative, from local governments that provide lodging and permit the use of public space, allowing local politicians to tout their support of rural constituencies, to cooperation from IMCINE, which facilitates the licensing of state-funded domestic films for distribution. By limiting the films on offer to domestic cinema, Cine Móvil ToTo justifies its alignment with IMCINE's goal of fomenting national cultural production, essentially operating as an arm of this entity by filling the space of cut programs like El Cine en tu Comunidad. In this sense, Cine Móvil ToTo is symptomatic of the privatization of cinema in Mexico, in that it is a nonprofit that has stepped in to redress the state's abandonment of the distribution and exhibition of domestic cultural production.

In addition to state support, Cine Móvil ToTo relies on corporate sponsorships to finance their tours. The website pitches benefits to potential corporate partners in a variety of ways that are recognizable under the rubric of green capitalism. First is the appeal to a form of image branding or green washing: "associate your brand with social responsibility . . . as a friend to the environment"; "make your brand the face of the United Nations' sustainable development objectives." Second, sponsorship provides a means of amplifying brand visibility in hard-to-reach areas: "Each tour generates 20,000 direct engagements and these are multiplied by press and social media coverage, resulting in 8 million impacts over the course of five months." And finally, corporate support is a way of obtaining market research: "On each tour we can conduct activities like product sampling and demonstrations, as well as surveys of the audience in each community."[37]

To illustrate how this works, let's look at a tour in 2021 that was sponsored by Mexico's largest beer maker, Grupo Modelo. As part of the sponsorship package, ToTo screened a sponsored educational short about responsible drinking practices prior to the main feature and conducted a survey with a set of adult attendees before and after they viewed the sponsored content to gauge brand favorability.[38] These sorts of sponsorships have allowed ToTo to expand their scope and conduct multiple tours a year. But they also demonstrate how experiments in alternative and postcarbon exhibition structures must negotiate with the realities of corporate funding in the context of limited state support. Grupo Modelo's relatively small investment in funding the ToTo tour reaps the company various rewards, including increased brand visibility and the opportunity to conduct focus groups with hard-to-reach publics; it also is a way to tout their support of national cultural production and their green credentials. The sponsorship allows Grupo Modelo to self-fashion their image in a way that we might describe as green corporate nationalism. In terms of actual carbon reduction, Grupo Modelo can have a much larger impact by creating internal changes to their production model. The company is making moves to do so by embracing wind power sourced from Puebla as of 2019 (albeit in partnership with Iberdrola, a wind developer that has been involved in land disputes with Indigenous communities in Oaxaca) and has plans to be totally reliant on renewables by 2025. However, Grupo Modelo and their export partner Constellation Brands have been slower to address their breweries' elevated usage of public water in drought-stricken areas like Mazatlán and Mexicali.

In this context, Grupo Modelo's partnership with Cine Móvil ToTo is illustrative of how green cultural initiatives strategically buoy brand visibility in ways that can be understood as greenwashing: the superficial support for carbon reduction without structural change. This example underscores how postcarbon cultural production cannot be romanticized as somehow existing outside of neoliberal dynamics in which cultural infrastructures are financed by capital. For this reason, I see Cine Móvil ToTo as an illustrative example of contemporary subjunctive aesthetic forms, which at once aspire to imagine the world otherwise (experimenting with building, in this case, a more environmentally and socially just model of cultural distribution) and are simultaneously subordinate to existing economic systems (neoliberal models of private financing, capital that is inextricably tied to extractivist exploitation).

The bicycle is at the center of Cine Móvil ToTo's pitch to green expanded access to Mexican cinema, as the website puts it, "a partir del simple pero poderoso pedaleo de una bicicleta" (through the simple but powerful pedaling of a bicycle.)[39] The initiative's imagery also orbits around the bicycle. In the logo, the two *o*'s of "ToTo" are rendered as bike wheels that abstractly connect to a projection bulb through a looping yellow cord (fig. 5.5). This discursive and visual framing foregrounds the bicycle as ToTo's central powering device.

Yet manually operated bicycles can only generate so much energy. Each stationary bike produces around 300 to 600 watts per hour, according to Rodrigo Soto, Cine Móvil ToTo's communication director. This output varies according to human capacity, error, and distractibility—variables that are particularly salient given the high involvement of enthusiastic youth. Because of these limitations, bikes are used to power the audio console, speakers, Blu-ray, and other equipment that does not require continuous high-volume energy. The projector, by contrast, has higher energy demands (400 to 500 watts) and is powered by solar panels to ensure its seamless operation. These solar panels are strapped to the team van, the ToToneta, where they charge during the trip, and are then placed near the screening venue to continue to charge in advance of the show.

The bicycle is an effective shorthand for fossil fuel alternatives. A commonplace cheap technology, the widespread use of bicycles in rural and poor communities proves that postcarbon modes of social organization already exist. To associate the bicycle with futurity and technological innovation,

FIGURE 5.5. The Cine Móvil ToTo logo features bicycle wheels abstractly connected to a projection bulb. Courtesy of Cine Móvil ToTo.

as Cine Móvil ToTo does in its branding, scrambles the typical collapse of modernity with high-carbon lifestyles or expensive green technologies that are nonetheless reliant on the extractivist exploitation of minerals, like lithium, an essential material for the batteries needed to power electric vehicles.[40] The insights of Marxist philosopher Bolívar Echeverría help elucidate this point. Echeverría has argued that the idea of modernity has become so subsumed by capitalism and whiteness that they appear coterminous. Yet modernity demands to be reimagined beyond its contingent capitalist iteration, which promises abundance but undercuts this promise through its exploitative relations of production.[41] Modernity, Echeverría urges, should be conceptualized as the potential of technological change to serve principles that stretch beyond individual accumulation, like collaboration and community. The bicycle is one such example of a technology that possesses this subjunctive potential to reorient modernity beyond accumulation. From this vantage, the widespread use of bicycles in rural communities is not an indicator of belated modernity, but a model of postcarbon living that articulates one path toward a planetary future.

While ToTo's visual and discursive focus on the bicycle reinforces the bicycle's association with postcarbon technologies, the greater efficacy of solar panels at meeting the high-volume energy requirements of film projection presents us with the question of why this technology is absent from ToTo's messaging. We might speculate that the bike is given greater weight than solar because pedal-powered machines invite community involvement in the production of green energy. Yet to stress the audience's role in reducing emissions when solar power is what truly undergirds ToTo screenings reinforces the narrative that the responsibility to reduce emissions lies with the individual. This is misleading on various fronts. Mexico itself is responsible for only about 1 percent of global emissions, and the low-income communities that benefit from Cine Móvil ToTo produce just one-seventh of the emissions of their wealthier counterparts in Mexico.[42] Thus, while the centrality of bikes in allowing individuals to power alternative exhibition infrastructures makes the energy demands of cultural consumption visible and corporeal, this visual shorthand also obscures the fact that the solutions to the energy crisis are collective and require large-scale systemic change.

FROM AMBULANT TO SELF-SUSTAINING EXHIBITION

While cinema as an industry is far from a key emitter when compared to sectors like transportation and manufacturing, Cine Móvil ToTo nonetheless illustrates how the subjunctive reorganization of cultural infrastructure to prioritize low emissions and equitable access can be put into action. This model is just the jumping-off point for some subjunctive speculation of our own. How might rural communities not just be visited by outside initiatives like Cine Móvil ToTo but also be equipped with solar infrastructure to generate their own projects and exhibition spaces?[43] Such a move would shift focus from the active involvement of rural participants in reducing their already insignificant energy consumption by pedaling and reorient it to providing communities with the technologies that will allow them to create their own postcarbon cultural infrastructures.

In that sense, Cine Móvil ToTo's green exhibition model might be brought into generative dialogue with efforts like Aquí Cine, a nonprofit network that supports community-led exhibition spaces. Founded in 2011 by Damián López and Luna Marán as an itinerant film initiative serving rural communities in Oaxaca, in 2013, Aquí Cine morphed into something different: a platform to help rural cinephiles establish more permanent exhibition sites in their communities. In this new iteration, Aquí Cine operates as the hub around which locally organized, autonomous spokes gravitate. The initiative works as follows: someone from a community that lacks an exhibition space contacts Aquí Cine, which helps them identify a space for an alternative exhibition site or *cineclub* and then provides guidance on how to arrange that space and obtain necessary equipment. Once the site is established as part of its network, Aquí Cine offers free film programming obtained from festivals, distributors, and filmmakers, as well as promotional materials to encourage attendance. Aquí Cine also puts on workshops to further train its community organizers and support the longevity of each spoke. Through these efforts, Aquí Cine has enabled the formation of seventeen alternative exhibition initiatives, predominantly in Indigenous (Zapotec, Mixe, Mixtec, Mazatec) communities in Oaxaca.[44]

Aquí Cine's transition from an ambulant theater into a distribution-exhibition infrastructure illustrates how the problem of access can be tackled in a way that is community driven and self-sustaining. While the ephemeral nature of the Cine Móvil Toto is buoyed by the excitement of a transitory experience, the spokes supported by Aquí Cine build community over time around the shared space of the theater. As Fernanda Río eloquently puts it in *Manual para exhibicionistas: cómo montar un cine* (Manual for Exhibitors: How to Establish a Movie Theater, 2015), a guide based on her experience with Aquí Cine, "el congregarse en los límites de la oscuridad, se vuelve un acto de complicidad, una situación común que convierte a los individuos en una mente que piensa en colectivo, o una mente colectiva que pone el cine como una situación común" (congregating in the darkness becomes an act of collusion, a shared situation that turns individuals into a mind that thinks as a collective, or a collective mind that construes cinema as a situation in common).[45] Perhaps a future iteration of Cine Móvil ToTo might expand in this direction, helping rural communities establish decarbonized cinema infrastructure that will endure in time and serve the interests of the community as it sees fit.

Although the green exhibition model deployed by Cine Móvil Toto was developed in response to scarcity—the absence of film exhibition spaces in rural communities—such a model would be most impactfully adopted, not by those whose carbon footprint is already minuscule, but by those who have a disproportionate carbon footprint. We see this at work, for instance, in Solar World Cinema, a Dutch initiative that approximates the blending of the models of Aquí Cine and Cine Móvil ToTo. Founded in 2009 as an expansion of a local project in solar screening, Solar World Cinema provides interested partners with a digital toolkit explaining how to establish solar-powered ambulant cinema, including guidance on equipment, advice on financing, and networking opportunities.[46] Because the project is aimed at democratizing cinema and expanding access, Solar World Cinema has partners in countries ranging from Nepal to Chile. But it presents an opening to think about how models in postcarbon cinema infrastructure might be adopted by choice in urban centers and by film festivals. This is not a call for austerity, but rather for the building of film infrastructures that are plentiful in ways that are also equitable.

CINEMA'S ECOLOGICAL FOOTPRINT

As I've pointed out, Cine Móvil Toto reduces the carbon footprint of populations whose emissions are already negligible. So while it is a fascinating bid for alternative exhibition formats organized around equity, democratic distribution, and energy justice, it is not necessarily a project with resounding ecological impact. If we want to think explicitly about reducing cinema's carbon footprint, perhaps we are looking in the wrong place by attending to exhibition infrastructure or even Mexican cinema in the first place. That is because the greatest source of emissions associated with cinema is on the production end and specifically originates in air travel. In what follows, I briefly explore the connection between film production, transportation, and labor, as it brings into view tensions between the imperative to green cinema and the aspiration to create film industries that are more equitable.

A 2020 report by albert, a British environmental organization that provides the TV and film industry with guidance on how to reduce carbon emissions, found that the average film with a budget of US$70 million generates approximately 2,840 tons of CO_2. Over half of these emissions are attributable to the various forms of transportation required to shuttle the cast, crew, and equipment to different shooting sites.[47] albert signaled that reducing air travel is the most expedient way to slash emissions, a theory that was proven during the COVID-19 pandemic. It observed that pandemic-related travel restrictions and a rise in remote work produced a significant drop in the emissions of the average British TV production, as filmmakers reduced the size of cast and crews, shifted to more local shooting sites, and minimized air travel.[48]

Part of the reason that travel constitutes such a high percentage of film production emissions is that over half of all Hollywood films are shot outside of Los Angeles, a phenomenon known as runaway productions. This outsourcing is motivated by cost-reduction strategies like the search for cheap, highly skilled labor, as well as tax incentives and subsidies offered by location sites. Runaway productions have an inherently larger carbon footprint, as they require more travel: approximately 110 individual round-trip flights and numerous containers of equipment shipped for a typical $75 million film.[49]

In addition to expending higher emissions, runaway productions exert pressure on international environmental and labor regulations. Countries competing to attract runaway productions do so by promising more permissive regulatory conditions. This has created a domino effect in which even California, known for having some of the United States' strictest environmental laws, reduced environmental regulations imposed on Hollywood productions to retain shooting in the state.[50] Runaway productions reproduce an extractive view of cultural production that is deeply embedded in colonial paradigms, utilizing foreign locales as sites to be mined for talent and resources and as the dumping ground for cinematic waste under the purported justification of strengthening local economies and transnational film industries.

Illustrative of how the ecological tensions surrounding runaway productions play out in Mexico is Baja Studios. This massive production facility was built in 1996 in the coastal town of Rosarito, Baja California, by Twentieth Century–Fox to film the megahit *Titanic*. Initially conceptualized as a temporary facility, Baja Studios was built in just one hundred days for $20 million. This was incomparably cheaper than constructing such a facility in Los Angeles, an option that was never considered because of budgetary constraints.[51] As entertainment industry reporter Paula Parisi explained, "Cheap labor and real estate were necessities" to meet *Titanic*'s studio budget benchmarks.[52] Baja Studios' forty acres of oceanfront property, extensive unobstructed views of the Pacific Ocean, five soundstages, and four massive tanks, including a gigantic infinity pool tank that could hold twenty million gallons of water, made it at the time Hollywood's premiere complex for filming water-related productions.

Sophia McClennen has reported in *Globalization and Latin American Cinema* that most locals embraced the *Titanic* shoot as a massive opportunity for the community and those involved in the Mexican film industry, which at the time was experiencing a downturn.[53] Yet the hasty construction of the Baja Studios facility was accompanied by significant environmental impacts, including pollution and extensive runoff from the chlorine-filled tanks. Local fishermen "claimed that the volume of the catch of some species in the area around the studios had declined by 50 percent" and that the local sea urchin population was decimated.[54] The studio denied responsibility, and affected locals were not compensated for their lost livelihoods. "Extreme development" projects like Baja Studios carry other risks, as Ben Goldsmith and Tom O'Regan explain. While they may offer employment or training opportunities, locals "have little financial or creative control" over these projects or the established infrastructures.[55] Another consequence is sustained wage depression to ensure the continued supply of a financially competitive labor market.

The imperative to cut carbon emissions when considered in isolation from other economic, social, and environmental factors thus offers only a partial vantage of how to make cinema more compatible with environmental justice. If analyzed purely through the metric of carbon emissions, one might come to the conclusion that Baja Studios is relatively ideal. Just a three-hour drive from Los Angeles, Baja Studios represents a way for Hollywood productions to lower the emissions that accompany long flights.[56] The highly skilled film crews located in Mexico represent a cheap local labor force that does not need to be transported to far-flung locations. The presence of the studio also slashes shipping in half for equipment that can be locally sourced and stored.

But while from the standpoint of carbon emissions Baja Studios represents a way to green Hollywood, it does so in a way that continues to outsource the social and environmental costs of cultural production south and ultimately does not trouble the unequal and wasteful regimes of cultural and energy production that have given shape to the planetary climate crisis. The example of Baja Studios signals the need to imagine forms of cultural production and cultural infrastructure that not only reduce emissions but that also do not perpetuate the colonial dynamics that define global cinema.

CINEMA AS IT COULD BE

The carbon footprint of the Mexican film industry is already incomparably lower than that of its counterparts in the United States and Europe because of the industry's lower budgets, smaller casts and crews, and lower reliance on distant location shoots. In this comparative sense, even a blockbuster of Mexican cinema like *Cindy la Regia* (directed by Catalina Aguilar Mastretta and Santiago Limón), the highest grossing Mexican film of 2020 and among the top thirty highest grossing domestic films of all time, already represents a model of green filmmaking, even without consciously aiming to be so. This is a somewhat silly claim by design. The film is an ode to consumerism, with no environmentalist content or messaging whatsoever. Nonetheless, it has a low carbon footprint relative to comparable Hollywood films. I make this fast-and-loose claim based on a series of assumptions, since *Cindy la Regia*, like most Mexican films, did not track emissions during filming. But by using information sourced from the aforementioned albert report, we can make an informed guess. albert shows that the larger a film's budget, cast, and travel needs, the larger its carbon footprint. A film like *Cindy*, which cost approximately $2 million to make and was filmed in only two locations, thus inadvertently and imperfectly models the best practices advocated for by albert and represents a model for popular, audience-resonating filmmaking that could be adopted by filmmakers throughout the globe, even those uninterested in environmentalist themes.

Of course, other filmmakers are approaching the task of greening Mexican film production with more rigor. Mexican filmmaker Mónica Álvarez Franco is a pioneer on this front with her documentary *Bosque de niebla* (The cloud forest, 2017), the first Mexican feature-length film to track and reduce its ecological impact. The documentary profiles the protection of the cloud forest by a community in Veracruz. Following the recommendations of the UN Climate Change initiative Climate Neutral Now, Álvarez Franco's crew measured their energy expenditures while filming, reduced them whenever possible, and bought carbon offset credits from the El Verde Landfill Gas Recovery and Flaring Project in León.[57] They also chose not to use artificial lighting or butane gas and recharged their equipment with solar power.

In a talk about green cinema practices at Los Cabos International Film Festival, Álvarez Franco explained that she wanted to explore how filmmaking might emulate and adopt the principles of permaculture modeled by the community in Veracruz that her film profiled:

> Para mí es muy importante entender los principios de la permacultura y intentar adecuarlos o transformarlos a la realidad del cine. [. . .] [E]stos

principios no solamente se basan en la ecología sino en el cuidado de otras personas, en tener relaciones laborales más horizontales. [...] Si pensamos en la simbiosis de lo ecológicamente responsable con lo creativo del cine desde las propuestas del argumento, desde la idea creativa, desde el inicio del todo, creo que será mucho más fácil y orgánico lograr un objetivo sin sacrificar ni la idea ni la ética.

For me it is very important to understand the principles of permaculture and to try to adapt or transform them to the realities of filmmaking. ... These principles are not just based on ecology but on the care for others, and on establishing more horizontal labor relations.... If we think about the symbiosis of what is ecologically responsible with the creative force of cinema from the level of plot, the creative idea, from the beginning, then it will be easier and more organic to achieve an (ecological) objective without sacrificing our ideas or our ethics.[58]

Álvarez Franco argues that filmmakers should not think about reducing emissions as a burden or as creatively limiting. Instead, she suggests that "ecologically responsible" filmmaking practices can lead to formally innovative outcomes. Crucially, she signals that approaching cinema through the lens of environmental justice is expansive in implication. It entails rethinking cinematic practice and aesthetics, as well as labor practices. The task of reimagining cinema in the era of climate change is capacious in its invitation to try out filmmaking practices that will engender a more just future, as well as one that is less reliant on fossil fuels.

The need to delink cinema from oil is increasingly evident. Mexico has experienced a steady decline in crude output, which has halved since 2004. The causes behind this decrease include diminishing reserves, outdated technologies, and neoliberal energy reform policies that have tipped the scales toward imports from foreign transnationals. As a result, government revenue has suffered: in 2019, oil accounted for only about 17 percent of federal revenue, in sharp contrast to the 40 percent that had been the norm for decades. This steep reduction has again translated to significant cuts to state institutions, social programs, environmental agencies, and arts initiatives. The film industry has been particularly affected: the total amount of state funds dedicated to cinema was cut in half between 2012 and 2017; further austerity measures in 2019 slashed the remaining film sector budget by an additional 40 percent.

As I have argued here, the crisis of oil compels the subjunctive reinvention of infrastructures that serve the commons, a commons that is collaboratively conceived, equitable, and durable in the long term. Attending to the "infrastructures of distribution," Christian Sandvig writes, reveals "competing ideas about what content and which audiences are valuable, and indeed how culture itself ought to work."[59] With this in mind, let us not romanticize what ecocinema can accomplish without also grappling with its structural limitations, circumscribed as they are by the long histories of colonial capitalism. What does it mean to make and consume audiovisual content about subjects who cannot access it themselves because of the lack of exhibition spaces, insufficient streaming bandwidth, or prohibitively expensive tickets? Are handheld devices the only answer to this dilemma? Or might cinema spectatorship be reimagined?

Questions like these point toward the importance of developing methods of scholarly inquiry that cultivate "the connections that inhere between media *about* the environment and media *in* the environment."[60] The web series *El tema*, the Cine Móvil ToTo screening initiative, Aquí Cine, Baja Studios, *Cindy la Regia*, and *Bosque de niebla* offer different entry points to these questions. They all indicate that climate is the issue to address. This can be accomplished through content but also through choices made in production, format, and distribution, whether it be the use of limited casts and location sites, the creation of short, easily shareable videos on YouTube, or the harnessing of the sun to power exhibition. These alternative models indicate that Mexican cinema is at the forefront of reimagining postcarbon cultural production. The search for alternative models of production and viewership is motivated, not by a belief in the moral superiority of austerity or scarcity, nor out of the nostalgia for declining forms of collective film watching, but rather by the forward-looking search for what cinema can do to build and sustain community in the context of its neoliberal erosion and how it can move away from the reliance on fossil fuels and help chart energy futures that prioritize social and environmental justice.

Conclusion

Anthropogenic climate change is but the latest manifestation of the "civilizational crisis" of colonial-capitalist extractivism diagnosed by Mexican environmental economist Enrique Leff in *Ecología y capital* in 1986.¹ Bolívar Echeverría has since elaborated that crisis is the system's telos, writing that "the goal repeatedly reached by the process of reproduction of wealth in its capitalist mode is genocidal and suicidal at the same time."² This is because the logic of accumulation is operationalized via the exploitation of land and labor to the point of its foreclosure, "so long as its destruction serves the always-converging interests of capitalist accumulation."³ Following the leads of Leff and Echeverría, who underscore that this destructive force is leveraged against both human and nonhuman life, in this book I have argued that studies of late capitalism in Mexico must take the nonhuman environment into account, both as an object of analysis as well as an analytical lens through which to understand the contingent and codependent relationship of territory with cultural practices, aesthetics, values, and politics.

Each chapter has explored a distinct iteration of potential foreclosure resulting from global and local practices of capitalist exploitation, including the long wake of extractivist toxicity, bare violence against land defenders, the sixth extinction, the precarization of rain-fed agriculture by global warming, and the erosion of cultural infrastructure. In response, the artists, writers, filmmakers, and cultural practitioners studied in this book sketch out rejoinders to foreclosure through renewed imaginative engagement with the world as it is and as it could be. These "subjunctive aesthetics" embrace fear, speculation, and yearning as modes of response that begin to work through the impasse of the extractivist model and imagine ways of living otherwise, even in contexts circumscribed by inherited damage.

Other scholars of Latin American culture have similarly grappled with what art can do in the face of rampant environmental degradation. It can induce trance, as Jens Andermann proposes, allowing the spectator to become reenchanted by nonhuman vibrancy, realigning the relationship between human and nonhuman bodies outside of the paradigm of utilitarian consumption.[4] Or it can toggle between local and planetary scales, as Rachel Price suggests, forging connections between the global circuits of capital and the local devastation of place.[5] It can also make visible what Macarena Gómez-Barris theorizes as "submerged perspectives," points of view that have been suppressed by the imperialist gaze.[6] *Subjunctive Aesthetics* is akin to such formulations in that it too is invested in how cultural production is able to challenge seemingly uniform perceptions of reality promoted by extractivist ideologies.

The subjunctive turn marks a shift in dominant tendencies in ecocultural production away from modes of response like nostalgia and melancholia, identified by Charlotte Rogers as ascendant in the twentieth century, as well as a divergence from the contemporaneous surge in forensic aesthetics focused on proving harm and the mimetic inclination, evaluated by Victoria Saramago, to preserve ecosystems through fiction.[7] In contrast with the archival, evidentiary, elegiac, or retrospective force of these narratives and aesthetic trends, subjunctive aesthetics reflect the uncertainty of this moment of political impasse in which the visible symptoms of climate change coexist with the systemic doubling down on the extractivist model. Rather than consolidate around a single retort or solution to this quagmire, subjunctive aesthetics lean into the provisional nature of this as a moment that necessitates debate, interpretation, and affective response. Like the subjunctive grammatical mood, subjunctive aesthetics are a space for the contradictory and multivalent expression of desires and suggestions, but also fears, doubts, and hypothetical speculations. While subjunctive aesthetics encompass speculative forms (nonmimetic fiction that departs from the real), I purposefully use the term *subjunctive* because it more expansively indexes the ability of cultural expression to gesture toward the possible as well as to think, feel, and doubt rather than inform or mandate in response to the ethical quandaries of the human relationship to the planet.

The subjoined structure of the subjunctive grammatical mood illustrates that this is a form of expression that is relational (mediating between two subjects) or reactive (responding to a given circumstance and wishing or doubting it otherwise). Whether relational or reactive, subjunctive aesthetics do not necessarily pitch in any given direction—utopian or dystopian—but reflexively acknowledge the work's emergence in relation and in subordination to other forms, be they planetary climatic conditions or capitalist structures of production. This subjoined structure characteristic of subjunctive aesthetics recognizes the limitations that the assemblage places on individual agency, constraints that are ecological and utopian in their approximation of how the world is cocreated, but also materialist and dystopian in their subordination to the world that we have inherited, defined as it is by coloniality and capitalist extractivism.

In the era of climate change, when doubt, anxiety, and hope jumble together to form the structure of feeling, the subjunctive becomes a powerful conceptual mode of response. It opens reality up to renegotiation: peddling possibilities rather than certainties. Literature, film, and visual art are uniquely capable of channeling subjunctive responses that do not just assert the facts of the crisis but that imagine hypothetical exits from the cul-de-sac of extractivist development. We see this at work in Chapter 1 in Verónica Gerber Bicecci's intensive praxis of rewriting, which approximates through literary form the sedimentation of human activities and discourse. While Gerber Bicecci mirrors extractivism's seemingly infinite reiteratibility in *La compañía*, the act of rewriting is itself subjunctive in its activation of contingent transformation. Such a method underscores that any text—or, for that matter, any territory or economic system—is not fixed, but can shift to give rise to new meaning. Similarly, the counterfactual mourning performed in response to the murder of land defenders by activists and artists like Naomi Rincón Gallardo, studied in Chapter 2, deploys negation and speculation to contest that death is a definitive end, undoing the intent of bare violence to eradicate ways of conceptualizing and inhabiting territory outside of profit. Inspired by Indigenous practices of pluriversal land defense, counterfactual practices of mourning affirm the futurity of territory, life, and cultural flourishing.

Subjunctive yearning is also at the core of Maricela Guerrero's poetics of extinction in *El sueño de toda célula*, propelling her search for another form of expression outside of the language of empire. The imaginative task of building another language is all encompassing: "No hay hora ni lugar ni espacio en que no / anduviera buscando un lenguaje hecho de manos y viento y nutrientes; en que no / estuviera investigando una forma redonda / y conveniente de nutrirlos / de acompañarlos" (There is no time or place or space in which I am not / looking for a language made of hands and wind and nutrients; in which I am not / searching for a form that is a perfect / and a convenient way to nourish them / accompany them).[8] For Guerrero, as well as the other poets I discuss in Chapter 3, contiguity is a formal literary property through which to bring different lifeforms into figurative proximity and relation, while retaining ontological difference. Poetry is also a means to experiment with expressive forms that aspire to transcend anthropocentrism: "quisiera hablar en árbol y cobijarles" (I would like to be able to speak in tree and blanket you).⁹ All of these subjunctive exercises and expressions are deliberately provisional, they center art as a space for working through political impasse by activating fears, rage, and desire to imagine other possible worlds.

At the same time, contemporary art does not exist outside of the extractivist economic system, and thus it would be a mistake to see it as wholly oppositional. Even those objects that inscribe postextractivist values and narratives must still negotiate the neoliberal funding structures and uneven dynamics of cultural circulation. Chapter 4 explored how this tension plays out in documentaries that celebrate rural resilience, like Betzabé García's *Los reyes del pueblo que no existe*, which profiles those who boldly chose to remain in San Marcos after it was flooded by the construction of the Picachos Dam. But the meaning of resilience morphs as these films take flight as commodities on the transnational festival circuit. Considering the dynamics of cultural circulation is imperative for scholars of ecocriticism and contemporary Mexico, as it allows us to question who these objects are for and how their consumption by different publics shifts their messaging. In Chapter 5, I presented infrastructure as another avenue for rethinking the relationship between cinema and extractivism. If the works examined in the first three chapters signal the imperative to rethink the relationship between territory and future, how might this be put into practice through cultural infrastructures? Cine Móvil ToTo represents an attempt to reimagine cinema exhibition for the era of climate change and simultaneously redress issues of access and energy. It signals that enacting cultural production oriented around ethical living in the era of climate crisis entails a multifaceted approach that includes territory as one of several components to consider in the task of reimagining a more just valuation of life.

The material conditions that delimit cultural production represent fertile terrain for future exploration and even collective action. Increasingly, artists who critique extractivism in their work are beginning to think about these questions at the level of praxis as well. We see this in Mónica Álvarez Franco's approach to filming *Bosque de niebla*, the first Mexican film to track its greenhouse gas emissions, inspired by the community in Veracruz that it profiles, mentioned in Chapter 5. We also see this in the reflexive nods made by Verónica Gerber Bicecci and Naomi Rincón Gallardo to the role of the FEMSA Foundation in financing their projects, *La compañía* and *Sangre pesada*, both of which were commissioned by the XIII FEMSA Biennial in 2018, "Nunca fuimos contemporáneos." The FEMSA Biennial, Mexico's most important contemporary art platform for emerging Mexican artists, is funded by the charitable arm of the multinational FEMSA corporation, the world's largest bottler of Coca-Cola products. Like other industrial bottlers (such as Grupo Modelo, discussed in the last chapter), FEMSA has been granted dozens of water concessions in Mexico. Its heavy use of public aquifers, from which a single FEMSA plant is permitted to extract between 300,000 to 1 million gallons of water a day, has contributed to ongoing water shortages in the communities where it operates, and has been met with protests, including a twenty-year effort to revoke the FEMSA water concession in San Cristóbal de las Casas.[10] FEMSA's exploitation of aquifers puts pressure on weak state infrastructure, and then profits from providing the solution to the problem it creates, stepping in to sell bottled water back to the community.[11] This situation has only worsened as climate change has made precipitation increasingly irregular, reducing the amount of water available to locals.

FEMSA's charitable arm, the FEMSA Foundation, has since 2008 funded "sustainability-oriented social investment projects," focused on water, childhood development, and culture—the Biennial chief among them.[12] The foundation's efforts buttress the corporation's purported commitment to sustainable water use, by, for example, funding The Latin American Water Funds Partnership, a self-described "innovative way to finance the protection and restoration of forests and grasslands surrounding the watershed [and] to help provide clean water to millions . . . across Latin America."[13] Such charitable efforts counter the impression that FEMSA is an irresponsible resource manager, and constitute a form of greenwashing through highly visible, non-systemic commitments to sustainability.

This points to an ethical conundrum at the heart of environmentally oriented cultural production like Gerber Bicecci's *La compañía* and Rincón Gallardo's *Sangre pesada*, which, in this light, participate however unavoidably in FEMSA's greenwashing or "artwashing," defined by Jasbir Puar and Andrew Ross as "the custom of using art and culture to launder ill-gotten gains and predatory practices . . . that stem from industries that harm the very communities that are supposed to enjoy and benefit from museums."[14] On the other hand, what better use for FEMSA's money than to critique extractivist practices, as Gerber Bicecci and Rincón Gallardo do in their pieces? This is particularly the case given the ever decreasing public arts funding in Mexico. And as we have seen, even public funding does not offer an out for those seeking financial backing disentangled from extractivism, given that 40 percent of Mexico's operating revenue has been historically derived from oil, and as such, there is a longstanding constitutive relationship between Mexican cultural industries and extractive enterprise.

In a conversation with me, Rincón Gallardo commented that although she referenced FEMSA in her anti-extractivist performance *Sangre pesada* through the inclusion of plastic Coca-Cola bottles, she would likely not participate in another FEMSA Biennial, given the ethical complexities of accepting its funding.[15] Gerber Bicecci similarly explained in an artist talk that her engagement with extractivism in *La compañía* and water contamination in the companion project *La máquina distópica*, "was a way of speaking about FEMSA without directly addressing FEMSA. At the same time, it was a way of confronting the situation that I was in, which was that my artistic project was being financed by a company that produces the same material damage on the landscape that I wanted to investigate."[16] This reflexive acknowledgment of the inextricability of art from the industries that fund it makes sense within Gerber Bicecci's stated goal to "depurify art and literature, to show that it isn't removed from extractivism or from the language of neoliberalism."[17] Thinking about these materialist entanglements alongside the content of the work opens up new possibilities for their interpretation. The fact that FEMSA funded the creation of *La compañía* allows us to read the titular company in a new light. Perhaps, in addition to alluding to Rosicler and other mining companies that have operated in Zacatecas, we can read the titular company as referring to FEMSA itself, a company whose presence the artist must deal with as an insidious invited guest, who must be negotiated with and perhaps eventually overthrown.

Gerber Bicecci's caution against demands for artistic purity is well warranted. Alexis Shotwell has made the case against purity politics as a political dead end, writing that "if we want a world with less suffering and more flourishing, it would be useful to perceive complexity and complicity as the constitutive situation of our lives, rather than as things we should avoid."[18] Rather than seeking a position that is somehow outside of the dynamics of exploitation inherent to late extractivist capitalism, we might instead recognize ourselves as implicated within its workings. This recognition of implication—akin to what I have described as the subjoined characteristic of subjunctive aesthetics—affords a more robust conceptualization of collective responsibility. As Michael Rothberg observes, "a theory of implication allows us to retain our sense that situations of conflict position us in morally and emotionally complex ways and yet still call out for forms of political engagement that cut through complexity to remain on the side of justice."[19]

Gerber Bicecci and Rincón Gallardo's anti-extractivist pieces commissioned for the XIII FEMSA Biennial dramatize these complexities in a way that is paradigmatic of what I call "subjunctive aesthetics." That is, artworks that are less interested in the evidentiary mode of making environmental damage visible, and more invested in the capacity of art to feel its way through the uncertainty and ambiguity of reiterative cycles of extractivism, activating at once the possibilities of imaginative thought, yet reckoning with its subjoined relationship to extant power structures of accumulation. Through the method of rewriting, Gerber Bicceci's *La compañía* models what it means to inherit a delimited space thoroughly shaped by centuries of extractivism, and yet, within that space, to animate contingency and potential transformation through interventions that are seemingly minor in scale.

Thinking about how art emerges from within extractivist infrastructures is an important step in reckoning collective cultural implication in extractivist economies and ideologies. The successful 2019 campaign by Indigenous and environmentalist activists in Oaxaca to oust Compañía Minera Cuzcatlán (an affiliate of the Canadian transnational Fortuna Silver Mines, mentioned in Chapter 2) as a sponsor of the Oaxaca FilmFest represents an exciting next step in this reckoning.[20] We might hope for continued reflexive engagement with the conditions of cultural production as part of the subjunctive quest to imagine a postextractivist world that recognizes, following Luna Marán, that "la tierra es eso que hace que seamos" (land is that which makes us be).[21]

Notes

INTRODUCTION

1. Luna Marán, "Piel territorio," Tzam Trece Semillas, July 2021, https://tzamtrece-semillas.org/sitio/piel-territorio.
2. Marán, "Piel territorio," my emphasis. The imbrication of land, self, and community is characteristic of Indigenous approaches throughout the Americas to territory as a weave or a commons that is "ongoing, always in the making." Mario Blaser and Marisol de la Cadena, "The Uncommons: An Introduction," *Anthropologica* 59, no. 2 (2017), 186.
3. Real Academia Española, s.v. "modo subjuntivo," https://dle.rae.es/modo#KSAI6F0.
4. John Butt and Carmen Benjamin, *A New Reference Grammar of Modern Spanish* (New York: Routledge, 2011), 241.
5. Caterina Mauri and Andrea Sansó, "The Linguistic Marking of (Ir)Realis and Subjunctive," *The Oxford Handbook of Modality and Mood*, ed. Jan Nuyts and Johan Van Der Auwera (Oxford, UK: Oxford University Press, 2016), 175.
6. Soren Kierkegaard, *Papers and Journals*, trans. Alastair Hannay (New York: Penguin Books, 2015). Dorothy Wang elaborates that the subjunctive is "the necessary condition of all writing," because the act of putting pen to paper reflects the creator's desire to question or remake the world. Dorothy Wang, *Thinking Its Presence: Form, Race, and Subjectivity in Contemporary Asian American Poetry* (Stanford, CA: Stanford University Press, 2014), 292.
7. Natalia Mendoza, "2014: Mancha naranja de trescientos kilómetros," in *1968–2018: Historia colectiva de medio siglo*, ed. Claudio Lomnitz (México: Universidad Nacional Autónoma de México, 2018), 476.
8. There are exciting developments in the use of realist techniques in contemporary Mexican literature to register ecological catastrophe even in works whose primary narrative aims lie elsewhere, like Cristina Rivera Garza's *El invencible verano de Liliana* (*Liliana's Invincible Summer*), which embeds observations

about chemical plants, water contamination, and the lack of green spaces as part of the author's drive to fully account for the territory where her sister lived before her murder. Cristina Rivera Garza, *El invencible verano de Liliana* (México: Random House, 2021), 30.

9. Marisol de la Cadena, "An Invitation to Live Together: Making the 'Complex We,'" *Environmental Humanities* 11, no. 2 (2019): 477.
10. Irmgard Emmelhainz, "Images Do Not Show: The Desire to See in the Anthropocene," in *Art in the Anthropocene: Encounters among Aesthetics, Politics, Environments, and Epistemologies*, ed. Heather Davis and Etienne Turpin (London: Open Humanities Press, 2015), 138.
11. Alejandra Amatto, "Transculturar el debate. Los desafíos de la crítica literaria latinoamericana actual en dos escritoras: Mariana Enriquez y Liliana Colanzi," *Valenciana* 13, no. 26 (2020): 215.
12. Samuel Delany, *The Jewel-Hinged Jaw: Notes on the Language of Science Fiction* (Middletown, CT: Wesleyan University Press, 2009), 65.
13. Amitav Ghosh, *The Great Derangement: Climate Change and the Unthinkable* (Chicago: University of Chicago Press, 2016), 128–29.
14. Anna Lowenhaupt Tsing, *The Mushroom at the End of the World: On the Possibility of Life in Capitalist Ruins* (Princeton, NJ: Princeton University Press, 2015), 3–4.
15. Andrés Bello's foundational 1847 definition of the subjunctive mood emphasized its subordinated character: "El subjuntivo comun tiene un carácter que lo diferencia de todo otro Modo, i que es subordinándose o pudiéndose subordinar a palabras o frases que expresan mandato, ruego, consejo, permisión, en una palabra, deseo (I lo mismo las ideas contrarias como disuasiones, desaprobación, prohibición), significa la cosa mandada, rogada, aconsejada, permitida, en una palabra, deseada (I la cosa disuadida, desaprobada, prohibida, etc) . . . Quiero/Deseo/Ruego/Te encargo/Permito/Te aconsejo/Te prohíbo—que estudies el derecho." (The common subjunctive has a characteristic that differentiates it from any other grammatical mode, which is that it is subordinated to words or phrases that express command, pleas, advice, permission, or in one word, desire [and the same with contrary ideas like discouragement, disapproval, prohibition], it means the thing that is required, requested, suggested, permitted, in one word, desired [and the thing that is dissuaded, disapproved, forbidden, etc] . . . I want/desire/ask/request/permit/recommend/prohibit—that you study law.) Andrés Bello, *Gramática de la lengua castellana destinada al uso de los americanos: Obras completas. Toma Cuatro* (Madrid: EDAF, 1984), 160.
16. Marán, "Piel territorio."
17. Ministry of Environment and Natural Resources, "Nationally Determined Contributions: 2020 Update," Government of Mexico, 2020, https://www4.unfccc.int/sites/ndcstaging/PublishedDocuments/Mexico%20First/NDC-Eng-Dec30.pdf.
18. Darcy Tetreault, "Extractive Policies in Mexico at the Outset of López Obrador's Presidency," in *Latin American Extractivism: Dependency, Resource Nationalism,*

and Resistance in Broad Perspective, ed. Steve Ellner (Lanham, MD: Rowman and Littlefield, 2020), 157.

19. Paula Serafini, *Creating Worlds Otherwise: Art, Collective Action, and (Post)extractivism* (Nashville, TN: Vanderbilt University Press, 2022); Macarena Gómez-Barris, *The Extractive Zone: Social Ecologies and Decolonial Perspectives* (Durham, NC: Duke University Press, 2017).
20. See Milton Santos, "The Return of the Territory," in *Milton Santos: A Pioneer in Critical Geography from the Global South*, ed. Lucas Melcaço and Carolyn Prouse (Cham, Switzerland: Springer, 2017), 25–31.
21. "En tiempos de crisis climática, el futuro es un territorio a defender," #FuturosIndigenas, June 2021, https://futurosindigenas.org/manifiesto.
22. Verónica Gerber Bicecci, Canek Zapata, and Carlos Bergen, "La máquina distópica," https://lamaquinadistopica.xyz.
23. Maricela Guerrero, *El sueño de toda célula* (Mexico: Antílope, 2018), 99.
24. Horacio Machado, *Minar: Colonialidad y genealogía del extractivismo*, 3rd ed. (Mexico: Ediciones OnA, 2020), 13.
25. Sivan Kartha et al., *The Carbon Inequality Era: An Assessment of the Global Distribution of Consumption Emissions among Individuals from 1990 to 2015 and Beyond* (Stockholm: Stockholm Environment Institute and Oxfam International, 2020), 7.
26. Douglas Starr, "Just 90 Companies Are to Blame for Most Climate Change, This 'Carbon Accountant' Says." *Science*, August 25, 2016. https://www.science.org/content/article/just-90-companies-are-blame-most-climate-change-carbon-accountant-says.
27. See Kyle Whyte, "Indigenous Climate Change Studies: Indigenizing Futures, Decolonizing the Anthropocene," *English Language Notes* 55, no. 1–2 (Fall 2017): 153.
28. Maristella Svampa, *Neo-Extractivism in Latin America: Socio-environmental Conflicts, the Territorial Turn, and New Political Narratives* (Cambridge, UK: Cambridge University Press, 2019), 15.
29. Thea Riofrancos, *Resource Radicals: From Petro-Nationalism to Post-Extractivism in Ecuador* (Durham, NC: Duke University Press, 2020), 6.
30. Germán Vergara, *Fueling Mexico: Energy and Environment, 1850–1950* (Cambridge, UK: Cambridge University Press, 2021), 2.
31. The myth of El Dorado, Charlotte Rogers has shown, established Latin America as a potent site for colonial speculation that hinged on the transformation of nonhuman nature into future riches, a promissory logic that continued after independence. Charlotte Rogers, *Mourning El Dorado: Literature and Extractivism in the Contemporary American Tropics* (Charlottesville: University of Virginia Press, 2019). Ericka Beckman has demonstrated how *modernista* writers drummed up "export reverie" for consumerist pleasures that could be brought into reach through the region's integration into the world economy as the provider of raw materials. Ericka Beckman, *Capital Fictions: The Literature of Latin America's Export Age* (Minneapolis: University of Minnesota Press, 2013), 5.

32. This is the case even after aggressive efforts to privatize Pemex since the nineties substantially weakened it in ways that were then used to further justify its restructuring. See Darcy Tetreault, "The New Extractivism in Mexico: Rent Redistribution and Resistance to Mining and Petroleum Activities," *World Development* 126 (2020): 6.
33. Claudio Lomnitz, "Ayotzinapa y la crisis de representación," in *La nación desdibujada: México en trece ensayos* (México: Malpaso, 2016), 49. Edith Negrín hypothesizes that oil's presence is so integral to nearly every aspect of modern life in Mexico, it has become naturalized, taken for granted. Edith Negrín, *Letras sobre un dios mineral: El petróleo mexicano en la narrativa* (México: El Colegio de México, 2017), 17.
34. Scarlett Lindero, "Pemex contaminó 655 lugares en México entre 2008 y 2021: Semarnat," *Gatopardo*, August 15, 2022.
35. Tetreault, "New Extractivism in Mexico," 5.
36. Bolívar Echeverría, *Modernity and "Whiteness,"* trans. Rodrigo Ferreira (Cambridge: Polity, 2019), 16.
37. Víctor M. Toledo, *Ecocidio en México: La batalla final es por la vida* (Mexico: Grijalbo, 2015), 15.
38. Echeverría, *Modernity and "Whiteness,"* 17.
39. Comisión Nacional de Hidrocarburos, "Análisis de Reservas de Hidrocarburos 1P, 2P y 3P. Al 1 de enero de 2021," April, 2021, https://www.gob.mx/cms/uploads/attachment/file/631695/2021.04.20._DSD_-_OdG_Reservas_al_1-ene-2021._vf-web-CNH.pdf, 26. Critics point to the massive amount of water required by fracking, which would strain water security in areas already affected by drought, as well as to its record of contaminating local water and air. Martha Pskowski, "Mexico's Fracking Impasse," NACLA, October 27, 2020, https://nacla.org/news/2020/10/22/mexico-fracking-impasse.
40. Manuel Llano Vázquez Prado and Carla Flores Lot, *La contribución de Pemex a la emergencia climática: Análisis de emisiones por campo petrolero desde 1960* (México: CartoCrítica, 2019).
41. Myrna Santiago observes, "Wherever PEMEX goes, illness, pollution, deforestation, degradation of land, elimination of ecosystems, destruction of wildlife, toxic neighborhoods, and acid rain follows." *The Ecology of Oil: Environment, Labor, and the Mexican Revolution, 1900–1938* (Cambridge, UK: Cambridge University Press), 354–55.
42. Echeverría, *Modernity and "Whiteness,"* 17.
43. Ejército Zapatista de Liberación Nacional, "Intervención de Marcos en la mesa 1 del Encuentro Intercontinental, 30 de julio de 1996," in *EZLN: Documentos y comunicados 3: 2 de octubre de 1995–24 de enero de 1997* (Mexico: Era, 1997), 323.
44. The on-the-ground antiextractivist work performed by countless communities throughout Mexico is being richly studied by scholars in Mexico. See Mina Lorena Navarro Trujillo, *Luchas por lo común: Antagonismo social contra el despojo capitalista de los bienes naturales en México* (Mexico: Bajo Tierra, 2015),

Griselda Sánchez, *Aire no te vendas: La lucha por el territorio desde las ondas* (Mexico: IWGIA, 2017), Felipe Reyes Escutia, ed., *Construir un NosOtros con la tierra: Voces latinoamericanos por la descolonización del pensamiento y la acción ambientales* (Chiapas: Universidad de Ciencias y Artes de Chiapas, 2018), Fausto Quintana Solórzano, ed., *Sociedad global, crisis ambiental y sistemas socio-ecológicos* (Mexico: UNAM, 2019).

45. Carlos Monsiváis, *"No sin nosotros": Los días del terremoto 1985–2005* (México: Era, 2005), 20. The relative invisibility of environmentalism and land defense in Mexico has extended to the cultural realm. In *Arte y olvido del terremoto*, Ignacio Padilla observes that the devastating 1985 earthquake was one of Mexico's three defining crises of the twentieth century, along with the Mexican Revolution and the 1968 Tlatelolco massacre. Yet whereas these other events were repeatedly thematized for decades thereafter, the earthquake barely registered. This willful cultural amnesia, Padilla speculates, is rooted in disenchantment: the sense that the mass civil mobilization that took place in the earthquake's wake did not ultimately lead to hoped-for political reforms. Yet the Revolution and Tlatelolco entailed similar disappointments and were nonetheless inscribed as watershed moments. So why did the earthquake not garner similar resonance? Perhaps the comparative silence surrounding the 1985 earthquake is symptomatic of a broader difficulty to discern the human role in seemingly arbitrary, natural phenomena. Ignacio Padilla, *Arte y olvido del terremoto* (México: Almadía, 2010).

46. Rocío Betzabeé González de Arce Arzave, "El viaje del cine mexicano de ficción hacia la conciencia ecológica: Imaginarios de la naturaleza, ecoutopías y ética ambiental en la pantalla" (MA thesis, Universidad Iberoamericana, 2019), 142.

47. Pablo Soler Frost, *Oriente de los insectos mexicanos*, 3rd ed. (Oaxaca: Zopilote Rey, 2019), 103.

48. Soler Frost, *Oriente de los insectos mexicanos*, 106.

49. Jill Carle, "Climate Change Seen as Top Global Threat," Pew Research Center, July 14, 2015, https://www.pewresearch.org/global/2015/07/14/climate-change-seen-as-top-global-threat.; Jaime Santos-Reyes, Tatiana Gouzeva, and Galdino Santos-Reyes, "Earthquake Risk Perception and Mexico City's Public Safety," *Procedia Engineering* 84 (2014): 662–71; Marshall Burke et al., "Higher Temperatures Increase Suicide Rates in the United States and Mexico," *Nature Climate Change* 8 (2018): 723–29.

50. Global Witness, "Standing Firm," September 2023, https://www.globalwitness.org/en/campaigns/environmental-activists/standing-firm.

51. Carolyn Fornoff, "Speculative Climate Change in Amado Nervo's 'Las nubes,'" *Paradoxa* 30 (2018): 15–34.

52. Gisella L. Carmona, *Ecologicon: Ecoliteratura en Alfonso Reyes* (Nuevo León: Universidad Autónoma de Nuevo Leon, 2012); Jorge Quintana-Navarrete, "José Vasconcelos's Plant Theory: The Life of Plants, Botanical Ethics, and the Cosmic Race," *Hispanic Review* 89, no. 1 (2021): 69–92; Kerstin Oloff, "The 'Monstrous Head' and the 'Mouth of Hell': The Gothic Ecologies of the 'Mexican Miracle,'"

in *Ecological Crisis and Cultural Representation in Latin America: Ecocritical Perspectives on Art, Film, and Literature*, ed. Mark Anderson and Zélia M. Bora (Lanham, MA: Lexington, 2016), 79–98; Victoria Saramago, "Juan Rulfo's *Pedro Páramo* and the Green Revolution: Modern Literary and Agricultural Dilemmas," in *Fictional Environments: Mimesis, Deforestation, and Development in Latin America* (Evanston, IL: Northwestern University Press, 2020), 93–122; Victoria Saramago, "Magueys and Machines: Narratives of Environmental Change in Mid-Twentieth-Century Mexico," *FORMA* 2, no. 1 (2023): 85–108; Mark D. Anderson, "Was the Mexican Revolution a Revolt of Nature?: Agustín Yáñez's Ecological Perspective," *Revista de Estudios Hispánicos* 40 (2006): 447–67; Laura Barbas-Rhoden, "Futuristic Narratives and the Crisis of Place," in *Ecological Imaginations in Latin American Fiction* (Gainesville: University of Florida Press, 2011); Brian Gollnick, "Ecology, Testimony, and the Politics of Representation," in *Reinventing the Lacandón: Subaltern Representations in the Rain Forest of Chiapas* (Tucson: University of Arizona Press, 2008), 121–46; Micah McKay, "'Pasto sin fin del basurero': Trash and Disposal in the Poetry of José Emilio Pacheco," *Latin American Literary Review* 47, no. 93 (2020): 49–58.

53. Homero Aridjis and Fernando Césarman, *Artistas e intelectuales sobre el ecocidio urbano* (Mexico: Consejo de la Crónica de la Ciudad de México, 1989); Carlos Fuentes, *Cristóbal Nonato* (Mexico: Fondo de Cultura Económica, 1988).

54. Founded in 1986 around the purported goal of "el cambio de actitudes en vías de . . . una sana relación con el medio ambiente" (changing attitudes toward a healthy relationship with the environment), the Partido Verde Ecologista de México (PVEM) has consistently aligned its platform with ultraconservative stances, including the embrace of the death penalty and privatizing the energy sector, to such an extent that PVEM was expelled from the Global Green coalition in 2011. PVEM's indifference to human rights and embrace of extractivist policies under the guise of green rhetoric has made a mockery of environmentalism, leading to understandable suspicion of conservationist discourse in Mexico. "Estatutos," Partido Verde Ecologista de México, 2010, https://www.partidoverde.org.mx/transparencia/II/Estatutos.pdf; Óscar Daniel Rodríguez Fuentes, "La lucha por la opinion pública en México: Sociedad civil vs. Partido Verde Ecologista de México," *Inciso* 18, no. 2 (2016): 21–35.

55. Gabriela Méndez Cota, "Policing the Environmental Conjecture: Structural Violence in Mexico and the National Assembly of the Environmentally Affected," *New Formations* 96/97 (2019): 82.

56. "Conoce estos espectaculares alebrijes en el Zoológico de Chapultepec," Wipy.tv, March 30, 2018, https://wipy.tv/alebrijes-en-el-zoologico-de-chapultepec.

57. Toledo, *Ecocidio en México*, 79.

58. Critics like Humberto Beck, Carlos Bravo Regidor, and Patrick Iber have noted that such pursuits, while well-intentioned, are powered by nostalgia for the stabilizing infrastructure boom of the mid-twentieth century and are out of touch with current concerns about environmental longevity and community input.

Humberto Beck, Carlos Bravo Regidor, and Patrick Iber, "Year One of AMLO's Mexico," *Dissent*, Winter 2020, https://www.dissentmagazine.org/article/year-one-of-amlos-mexico. AMLO made this comment in response to the murder of Samir Flores Soberanes. Darcy Tetreault, "Extractive Policies in Mexico at the Outset of López Obrador's Presidency," in *Latin American Extractivism: Dependency, Resource Nationalism, and Resistance in Broad Perspective*, ed. Steve Ellner (Lanham, MD: Rowman and Littlefield, 2020), 157.

59. Fernando Coronil, *The Magical State: Nature, Money, and Modernity in Venezuela* (Chicago: University of Chicago Press, 1997), 29.
60. Francisco Serratos argues that Jason Moore's term "Capitalocene" more accurately indexes the causes behind the human destabilization of the Earth system. I agree with this critique, but still find the terminology Anthropocene useful because of its wide adoption and legibility across disciplines. Francisco Serratos, *El capitaloceno: Una historia radical de la crisis climática* (Mexico: Festina, 2020).
61. Mary Louise Pratt, *Imperial Eyes: Travel Writing and Transculturation* (London: Routledge, 1992), 112.
62. Ghosh, *The Great Derangement*, 62–63.
63. Simon L. Lewis and Mark A. Maslin, "Defining the Anthropocene," *Nature* 519 (2015): 176. Nicholas Mirzoeff renames the Anthropocene "the White Supremacy Scene": a term that foregrounds the role of racism, colonialism, and empire at the core of the human transformation of the planet. Nicholas Mirzoeff, "It's Not the Anthropocene, It's the White Supremacy Scene; or, The Geological Color Line," in *After Extinction*, ed. Richard Grusin (Minneapolis: University of Minnesota Press, 2018), 123–49.
64. Alexander Koch et al., "Earth System Impacts of the European Arrival and Great Dying in the Americas after 1492," *Quaternary Science Reviews* 201 (2019): 13–36.
65. Kathryn Yusoff, *A Billion Black Anthropocenes or None* (Minneapolis: University of Minnesota Press, 2018), 33.
66. Alfred W. Crosby, *Ecological Imperialism: The Biological Expansion of Europe, 900–1900*, 2nd ed. (Cambridge, UK: Cambridge University Press, 2015).
67. Ivonne del Valle, "From José de Acosta to the Enlightenment: Barbarians, Climate Change, and (Colonial) Technology as the End of History," *Eighteenth Century* 54, no. 4 (2013): 435–59. See also Matthew Vitz, *A City on a Lake: Urban Political Ecology and the Growth of Mexico City* (Durham, NC: Duke University Press, 2018).
68. Raymond Williams, *Marxism and Literature* (Oxford, UK: Oxford University Press, 1977), 133.
69. Barbie Zelizer, *About to Die: How News Images Move the Public* (Oxford, UK: Oxford University Press, 2010), 15.
70. Anthony Giddens, *Modernity and Self Identity: Self and Society in the Late Modern Age* (Cambridge, UK: Polity, 1991).
71. Claudio Lomnitz, "Afterword: Spread It Around!" in *Cultural Agency in the Americas*, ed. Doris Sommer (Durham, NC: Duke University Press, 2006), 338.

72. Wai Chee Dimock, *Weak Planet: Literature and Assisted Survival* (Chicago: University of Chicago Press, 2020), 8.
73. Dimock, *Weak Planet*.
74. "The Anthropocene in Chile: Toward a New Pact of Coexistence," *Environmental Humanities* 11, no. 2 (2019): 468.
75. Mark Fisher, *Ghosts of My Life: Writings on Depression, Hauntology, and Lost Futures* (Winchester, UK: Zero, 2004), 36.
76. Dominic Boyer, *Energopolitics: Wind and Power in the Anthropocene* (Durham, NC: Duke University Press, 2019), 21.

CHAPTER 1

1. The centuries-long undertaking of the valley's desiccation, Reyes claimed, proved the State's resilience: "Son aquí ejemplo de cómo crece y se corrige la obra del Estado, ante las mismas amenazas de la naturaleza y la misma tierra que cavar" (They exemplify how the work of the State grows and corrects itself in response to the threats of nature and the Earth itself). Alfonso Reyes, *Visión de Anáhuac (1519)*, in *Obras completas de Alfonso Reyes II* (México: Fondo de Cultura Económica, 1995), 15. Robert T. Conn notes that Reyes interpreted the valley's draining as a "teleological process," a progressive historical endeavor. Robert T. Conn, *The Politics of Philology: Alfonso Reyes and the Invention of the Latin American Literary Tradition* (Lewisburg, PA: Bucknell University Press, 2002), 117.

    When Reyes first wrote *Visión*, he saw only the good that could come from the human manipulation of nature—manifest through the successive draining of Lake Texcoco—and not the unintended consequences of such an endeavor. Likewise, just as the term *Anthropocene* incorrectly suggests, with the root *anthropo-*, that all humans bear responsibility for climate crisis, in his eagerness to highlight the continuities between precolonial and colonial ecological transformation, Reyes glossed over radical differences. Whereas the Mexica who originally settled the area adopted a semiaquatic way of life, deploying a series of complex technologies that made the waterscape habitable, the Spanish imported and imposed their own ideas about what urban spaces should look like. They leveled Tenochtitlan, the city the Mexica had built on an island in Lake Texcoco and began a desiccation process that set off centuries of vexed water management in Mexico City, manifest through flooding, dust storms, water shortages, air pollution, and increased seismic vulnerability. See Ivonne del Valle, "On Shaky Ground: Hydraulics, State Formation, and Colonialism in Sixteenth-Century Mexico," *Hispanic Review* 77, no. 2 (2009): 197–220; and Vitz, *A City on a Lake*.

2. Reyes first coined the phrase "Viajero: has llegado a la región más transparente del aire" in his 1911 essay "El paisaje en la poesía mexicana del siglo XIX" and used it as an epigraph in *Visión de Anáhuac (1519)*, published four years later. Reyes, *Visión*, 13.

3. Quoted in Conn, *The Politics of Philology*, 117.
4. Natalia Soto-Coloballes, "The Development of Air Pollution in Mexico City," *Journal of Environment and Development* 10, no. 2 (2020): 4.
5. Alfonso Reyes, "Palinodia del polvo," *Caelum* 6 (2014): 62.
6. Reyes, "Palinodia," 64.
7. Pedro Ocampo Ramirez, "Pulso de México," *Jueves de Excelsior*, October 31, 1968.
8. Gisela Heffes gets at this idea from a different angle: the material practices of recycling by *cartoneros* and *pepenadores*. These cultural producers rewrite the detritus of capitalist regimes of consumption and waste. Gisela Heffes, *Políticas de la destrucción / Poéticas de la preservación: Apuntes para una lectura (eco)crítica del medio ambience en América Latina* (Rosario: Beatriz Viterbo, 2013), 243.
9. Lesley Wylie, *Colonial Tropes and Postcolonial Tricks: Rewriting the Tropics in the* Novela de la Selva (Liverpool, UK: Liverpool University Press, 2009), 7.
10. Wylie, *Colonial Tropes*, 15.
11. Cristina Rivera Garza, *The Restless Dead: Necrowriting and Disappropriation*, trans. Robin Myers (Nashville, TN: Vanderbilt University Press, 2020), 65. Estefanía Bournot also identifies resonance between *La compañía* and Rivera Garza's theorization of "geological writings" in Estefanía Bournot, "Abrir las heridas. Gerber, Meruane y Mendieta: Geoescrituras de un planeta enfermo," *Revista Letral* 25 (2021): 54–73.
12. Christian Moraru, *Rewriting: Postmodern Narrative and Cultural Critique in the Age of Cloning* (Albany: SUNY University Press, 2001), 9.
13. Roland Barthes, *Image-Music-Text*, trans. Stephen Heath (New York: Hill and Wnag, 1977), 148; Julia Kristeva, "Word, Dialogue and Novel," in *The Kristeva Reader*, ed. Toril Moi (New York: Columbia University Press, 1986), 36.
14. Moraru, *Rewriting*, 9. In Mexican literature, this sort of postmodern retelling is exemplified by Carmen Boullosa's *Son vacas, somos puercos* (They're cows, we're pigs, 1991), which borrows the structure and plot of seventeenth-century filibuster Alexander Olivier Exquemelin's memoirs, but retools them to foreground the Buccaneers' racialized violence, obviated in the original text. It also applies to the sort of oppositional occupation of neoliberal discourse performed by Hugo García Manríquez's *Anti-Humboldt* (2015), which replicates the North American Free Trade Agreement, and through selective erasure transforms it into a work of poetry by "refract[ing] the reader's gaze to the political conditions that surround [legal discourse]." Refurbishing is an apt metaphor for the type of intensive, revisionary rewriting undertaken by authors like Boullosa and García Manríquez, because it refers through the Old High German root *furben* to the parallel practices of cleaning and purging.
15. Walter Benjamin, *The Arcades Project*, trans. Howard Eiland and Kevin McLaughlin (Cambridge, MA: Belknap Press of Harvard University Press, 1999), 471. My understanding of Benjamin's approach as "a tradition of discontinuity"

is indebted to Susan Buck-Morss, *The Dialectics of Seeing: Walter Benjamin and the Arcades Project* (Cambridge, MA: MIT Press, 1989), 290.

16. Benjamin, *The Arcades Project*, 392.
17. Elizabeth DeLoughrey has noted that Benjamin's dialectical approach is useful for thinking about the Anthropocene because present manifestations of climate crisis, like sea-level rise or intensified drought, function as "a critical hermeneutic for reading the past." Elizabeth M. DeLoughrey, *Allegories of the Anthropocene* (Durham, NC: Duke University Press, 2019), 14.
18. Verónica Gerber Bicecci, *Empty Set*, trans. Christina MacSweeney (Minneapolis, MN: Coffee House Press, 2018), 82.
19. Gerber Bicecci, *Empty Set*, 83.
20. "Six Mexican Artists Revisit José Juan Tablada and His New York Circle," Proxyco Gallery (website), November 15, 2017, https://www.proxycogallery.com/talon-rouge.
21. Verónica Gerber Bicecci, *In the Eye of Bambi* (London: Whitechapel Gallery, 2020).
22. Verónica Gerber Bicecci, *La tierra es plana como una hoja (y cabalga en el aire)* (Mexico City: Gato Negro, 2021), https://www.veronicagerberbicecci.net/la-tierra-es-plana-earth-is-flat.
23. Geoff Bendeck, "In the Absence of Words: An Interview with Verónica Gerber Bicecci," Words without Borders, February 21, 2018, https://www.wordswithoutborders.org/dispatches/article/in-the-absence-of-words-an-interview-with-veronica-Gerber-bicecci.
24. Eduardo Ledesma, *Radical Poetry: Aesthetics, Politics, Technology, and the Ibero-American Avant-Gardes, 1900–2015* (Albany, NY: SUNY University Press, 2016), 171.
25. Willard Bohn, "The Visual Trajectory of Jose Juan Tablada," *Hispanic Review* 69, no. 2 (2001): 191–208, https://doi.org/10.2307/3247038.
26. Perla Velázquez, "Entrevista Verónica Gerber Bicecci, homenajes a Tablada y a Ampáro Dávila," *Bitácora de vuelos*, January 16, 2020, http://www.rdbitacoradevuelos.com.mx/2020/01/entrevista-veronica-Gerber-bicecci.html
27. Troubled by revolutionary politics, Tablada sided with the status quo, supporting Porfirio Díaz and his ideological successor, Victoriano Huerta. He pilloried revolutionary leaders in the press and satirized Francisco Madero as shamelessly power hungry in the never-staged play *Madero-Chantecler*, published in 1910 under a pseudonym. Rubén Lozano Herrera, *Las veras y las burlas de José Juan Tablada* (México: Universidad Iberoamericana, 1995), 224.
28. As Laura Torres-Rodríguez concludes, Japanese culture validated Tablada's belief that popular culture, if properly interpreted, could sustain the avant-garde search for an autochthonous yet modern national art. Laura J. Torres-Rodríguez, *Orientaciones transpacíficas: La modernidad mexicana y el espectro de Asia* (Chapel Hill: University of North Carolina Press, 2019), 51, 64.
29. Mauricio Tenorio-Trillo, *I Speak of the City: Mexico City at the Turn of the Twentieth Century* (Chicago: University of Chicago Press, 2012), 217.

30. Tenorio-Trillo, *I Speak of the City*, 222.
31. Torres-Rodríguez, *Orientaciones transpacíficas*, 60.
32. Tablada's *Un día* . . . not only brought the haiku to Mexico, it can also be considered Mexico's inaugural bestiary. In the 1970s, scholars Margaret Mason and Yulan Washburn described it as a "beast book" and a precursor of Juan José Arreola's reinvention of the bestiary. Margaret L. Mason and Yulan M. Washburn, "The Bestiary in Contemporary Spanish American Literature," *Revista de Estudios Hispánicos* 8, no. 2 (1974): 195.
33. Seiko Ota, "José Juan Tablada: La influencia del haikú japonés en *Un día* . . ." *Literatura Mexicana*, 16, no. 1 (2005): 139.
34. Ota, "José Juan Tablada," 135.
35. José Juan Tablada, *Un día . . . (poemas sintéticos)* (Caracas: [Imprenta Bolívar], 1919), Available online at "Juan José Tablada: Vida, letre e imagen," UNAM, 2021, https://www.iifl.unam.mx/tablada.
36. Ota, "José Juan Tablada," 143.
37. Verónica Gerber Bicecci, *Otro día . . . (poemas sintéticos)* (México: Almadía, 2019).
38. Gerber Bicecci, *Otro día* . . . .
39. Roland Barthes, *The Preparation of the Novel: Lecture Courses and Seminars at the College de France (1978-1979 and 1979-1980)*, trans. Kate Briggs (New York: Columbia University Press, 2011), 51.
40. Barthes, *The Preparation of the Novel*, 52.
41. Betina Keizman, "Territorios y naturaleza bajo la transmutación del archivo," *Valenciana* 12, no. 24 (2019): 229–45, https://doi.org/10.15174/rv.v0i24.473, 235.
42. "Voyager—What's on the Golden Record," NASA (website), accessed August 8, 2021, https://voyager.jpl.nasa.gov/golden-record/whats-on-the-record.
43. Lecia Rosenthal, *Mourning Modernism: Literature, Catastrophe, and the Politics of Consolation* (New York: Fordham University Press, 2011), 89.
44. Nuria Angélica Sánchez Matías, "Del desastre a los futuros habitables: La visualidad, la medialidad y la textualidad en la obra ecosocial de Verónica Gerber Bicecci" (MA thesis, Universidad Iberoamericana, 2023). See also Marta Pascua Canelo, "La mirada borrosa: Poéticas del desenfoque y visiones oblicuas en la narrativa hispánica contemporánea," *Catedral Tomada* 7, no. 13 (2019).
45. Gerber Bicecci, *Otro día* . . . .
46. Álvaro García Hernández, "El pueblo fantasma de Nuevo Mercurio y la minería insostenible," *Jornada Zacatecas*, July 9, 2014, http://ljz.mx/2014/07/09/el-pueblo-fantasma-de-nuevo-mercurio-y-la-mineria-insostenible.
47. The structural link between toxic masculinity and extractivism is discussed in David Loría Araujo and Francisco G. Tijerina Martínez, "La crisis del capital en dos energoficciones contemporáneas: *Temporada de huracanes*, de Fernanda Melchor y *La compañía*, de Verónica Gerber Bicecci," *De Raíz Diversa* 9, no. 17 (2022): 121–47. See also Gisela Heffes, "Toxicity," in *Handbook of Latin American Environmental Aesthetics*, ed. Jens Andermann, Gabriel Giorgi, and Victoria Saramago (Berlin: De Gruyter, 2023), 395–420.

La compañía can be situated within the burgeoning genre of Latin American ecohorror: environmentalist fiction that stages nature's revenge. Nicolás Campisi has identified other books that fit this mold, including Samanta Schweblin's *Distancia de rescate* (Argentina, 2014), Cristian Romero's *Después de la ira* (Colombia, 2018), Gabriela Alemán's *Poso Wells* (Ecuador, 2007), and Rita Indiana's *La mucama de Omicunlé* (Dominican Republic, 2015). Typically, nonhuman nature is the source of danger in ecohorror, but only because it is made so by human encroachment. Creature films from the 1970s, like the campy flick *Las abejas* (Alfredo Zacarías, 1978), dramatize nature fighting back against anthropogenic environmental harm. The psychological horror of these fictions derives from the realization that although we can change nature, we cannot control it—as in Reyes's "Palinodia del polvo." Simon Estok cautions, however, that the genre is ecophobic: founded on fear-based antagonism toward the nonhuman. In *La compañía*, the source of horror is not the land, but industry. This twist on the conventions of ecohorror signals that danger lies not in nature's potential revenge, but in the extractive paradigm that has produced a toxic landscape that is equally damaging for human and nonhuman life. Nicolás Campisi, "Ruinas contemporáneas: Ficciones del eco-horror en América Latina," *Afuera*, June 4, 2020, https://afuerablog.com/2020/06/04/ruinas-contemporaneas-ficciones-del-eco-horror-en-america-latina/; Simon C. Estok, "Theorising the EcoGothic," *Gothic Nature* 1 (2019): 48–49.

48. A wave of Spanish and Indigenous people migrated to Zacatecas after the discovery of silver. The existing Zacateco Indigenous population declined in the sixteenth century due to sickness and overwork brought by the mining economy. Valentina Garza and Juan Manuel Pérez, "La provincia minera de Zacatecas y su evolución demográfica (1700–1810)," *Historias* 77 (2010): 53–86.

49. Garza and Pérez, "La provincia minera," 57, 64.

50. Darcy Tetreault, "Water in Zacatecas: A Crisis without Conflict," in *Social Environmental Conflicts in Mexico: Resistance to Dispossession and Alternatives from Below*, ed. Darcy Tetreault, Cindy McCullough, and Carlos Lucio (New York: Palgrave Macmillan, 2018), 188.

51. See Bolívar Echeverría on how ownership of extractive technology is more politically influential than ownership of deposits, which leads to "a re-feudalization of economic life and the emergence of a transnational quasi-state." Echeverría, *Modernity and "Whiteness,"* 25.

52. Darcy Tetreault notes that mining booms and busts consolidated Zacatecas's role "as a territory for producing and exporting cheap labor." Tetreault, "Water in Zacatecas," 188.

53. Darcy Tetreault, "Free-Market Mining in Mexico," *Critical Sociology* 42, no. 4–5 (2016): 645. Neoliberal reforms that have accompanied the commodities boom of the twenty-first century have once again expanded mining in Zacatecas. Under Enrique Peña Nieto and Andrés Manuel López Obrador, Mexico

has adopted the new extractivist model forged by other Latin American progressive administrations. Neo-extractivism, according to Maristella Svampa, diverges from previous models in that the state actively captures and retains a greater slice of the profits through taxes, royalties, and rent. Surpluses are directed toward domestic infrastructure projects and social programs. This shores up public support for extractivism. But these progressive policies fail to address the concerns of rural communities and Indigenous populations that live in the extractive zone. Instead, they come with militarization and repression. Maristella Svampa, *Neo-extractivism in Latin America*, 6; Tetreault, "New Extractivism in Mexico," 2.

54. Víctor M. Toledo, David Garrido, and Narciso Barrera-Basols, "The Struggle for Life: Socio-environmental Conflicts in Mexico," trans. Mariana Ortega Breña, *Latin American Perspectives* 42, no. 5 (June 2015): 140.

55. Rocio Gomez, *Silver Veins, Dusty Lungs: Mining, Water, and Public Health in Zacatecas, 1835–1946* (Lincoln: University of Nebraska Press, 2020), 14.

56. Daviken Studnicki-Gizbert, "Exhausting the Sierra Madre: Mining Ecologies in Mexico over the Longue Durée," in *Mining North America: An Environmental History since 1522*, ed. J. R. McNeill and George Vrtis (Berkeley: University of California Press, 2017), 20.

57. Darcy Tetreault and Cindy McCulligh, "Water Grabbing via Institutionalized Corruption in Zacatecas, Mexico," *Water Alternatives* 11, no. 3 (2018): 576. More than 70 percent of all toxic emissions in Mexico are attributable to mining.

58. Gian C. Delgado, *Ecología política de la minería en América Latina* (Mexico: CIIH/UNAM, 2010).

59. Bruno Latour, *We Have Never Been Modern*, trans. Catherine Porter (Cambridge, MA: Harvard University Press, 1993). Marcela Romero Rivera similarly argues that Gerber Bicecci's method of intensive rewriting enacts a spectral temporal logic, in which the ghosts of colonialism and the period of the so-called "Mexican Miracle" persistently haunt the present. Marcela Romero Rivera, "Signs of the Inhuman: Hauntings and Lost Futures in Verónica Gerber Bicecci's *La Compañía*," *CLCWeb: Comparative Literature and Culture* 24, no. 1 (2022), https://doi.org/10.7771/1481-4374.4296.

60. As discussed in the conclusion, the FEMSA Biennial is considered the most important visual arts competition in Mexico. The FEMSA Biennial has been the site where some of the most interesting works about environmental issues have emerged, demonstrating the biennial's central role in funding and fomenting arts that respond critically and imaginatively to extractivism in Mexico. But the role of the FEMSA Biennial in the recent boom in Mexican environmental arts also points to the ethical complexity of contemporary arts funding. The FEMSA Foundation, which runs the biennial, is the charitable arm of the multinational FEMSA corporation, which is the world's largest bottler of Coca-Cola products, the second largest shareholder of Heineken, and the owner of OXXO convenience stores. Like other industrial bottlers, FEMSA has been granted

dozens of generous water concessions in Mexico; its heavy use of public aquifers (from which a single FEMSA plant is permitted to extract from 300,000 to 1 million gallons of water a day) has contributed to ongoing water shortages in the communities where it operates and has been met with community protests, including a twenty-year effort to revoke the FEMSA water concession in San Cristóbal de las Casas. Gerber Bicecci's *La compañía*, in this light, participates however unavoidably in FEMSA's greenwashing or artwashing. In an artist talk, Gerber Bicecci explained that her engagement with extractivism in *La compañía* and with water contamination in the companion project, *La máquina distópica*, "was a way of speaking about FEMSA without directly addressing FEMSA. At the same time, it was a way of confronting the situation that I was in, which was that my artistic project was being financed by a company that produces the same material damage on the landscape that I wanted to investigate." With this in mind, we might read FEMSA as another Company to which *La compañía* refers. "Verónica Gerber Bicecci: *La máquina distópica*," Seminario Investigación Poéticas de lo Inquietante, January 25, 2021, https://www.youtube.com/watch?v=tZtjCLTodTo&t=5845s.

61. *La máquina distópica, Oráculo web*, created by Verónica Gerber Bicecci, Canek Zapata, and Carlos Bergen, 2018, Toda la Teoria del Universo, accessed October 11, 2023, https://todalateoriadeluniverso.org/la-maquina-distopica-oraculo-web.

62. Alicia Sandoval, "*Otro día . . . (poemas sintéticos)* y *La compañía* de Verónica Gerber Bicecci," *Revista de la Universidad de México* 855/856 (Jan. 2020): 151. In his comprehensive analysis of *La compañía*, Nicolás Campisi writes that Part B models the tenets of collective social organization called *tequio* by Mixe theorist Floriberto Díaz. Nicolás Campisi, "Documentary Mines: Archives of Ecohorror in the Anthropocene," in *Post-Global Aesthetics*, eds. Benjamin Loy and Gesine Müller (Berlin: De Gruyter, 2023), 142.

63. Emily Hind, *Dude Lit: Mexican Men Writing and Performing Competence, 1955–2012* (Tucson: University of Arizona Press, 2019).

64. Ignacio Sánchez Prado, *Strategic Occidentalism: On Mexican Fiction, the Neoliberal Book Market, and the Question of World Literature* (Evanston, IL: Northwestern University Press, 2018), 177.

65. Marisol Luna Chávez and Víctor Díaz Arciniega, "La rutina doméstica como figuración siniestra. Amparo Dávila: Su poética del dolor," *Sincronía* 74 (2018): 205–33. Horror is not a dominant genre in Mexican literature. In the twentieth century, in addition to Dávila, writers like Manuel José Othón, Francisco Tario, and Guadalupe Dueñas dabbled in horror. So did canonical works like Carlos Fuentes's *Aura* and Juan Rulfo's *Pedro Páramo*. The absence of sustained literary engagement with horror in Mexico, author Rodolfo JM hypothesizes, can be attributed to the stigma attached to genre fiction, as well as to institutionalized ties to *costumbrismo*. Horror has been much more at home in Mexican film: taking off in the late sixties with the work of filmmakers like Carlos Enrique Taboada, and accelerating as a commercial genre in twenty-first

century, propelled by the international success of Guillermo del Toro and the domestic box-office hit *Kilómetro 31* in 2007. Literary horror has experienced a recent renaissance. Writers include Bernardo Esquinca, Fernanda Melchor, José Luis Zárate, Carlos Bustos, Andrés Acosta, Norma Lazo, FG Haghenbeck, Mario González Suárez, Luis Jorge Boone, Federico Vite, Antonio Malpica, Diego Velázquez Betancourt, Rodolfo JM, among others. Nonetheless horror remains relatively marginalized. Iván Farías puts it bluntly, explaining that those who write horror in Mexico do so in the face of "un aparato burocrático que privilegia el realismo" (a bureaucratic apparatus that privileges realism). Rodolfo JM, "13 ideas acerca de la literatura de terror mexicana," LJA.mx, May 27, 2012. https://www.lja.mx/2012/05/literatura-de-terror-en-mexico; Iván Farías, "Algunas precisiones sobre el terror en México," LJA.mx, May 27, 2012, https://www.lja.mx/2012/05/literatura-de-terror-en-mexico.

66. Amparo Dávila, *Cuentos reunidos* (México: Fondo de Cultura Económica 2009). Amparo Dávila, *The Houseguest and Other Stories*, trans. Audrey Harris and Matthew Gleeson (New York: New Directions, 2018).
67. Quoted in Erica Frouman-Smith, "Patterns of Female Entrapment and Escape in Three Short Stories by Amparo Dávila," *Chasqui* 18, no. 2 (1989), 50.
68. Frouman-Smith, "Patterns of Female Entrapment," 50.
69. Amparo Dávila, "Espejo lento," in *Poesía reunida* (México: Fondo de Cultura Económica, 2013), 46.
70. Eric Pennington, "Amparo Dávila's 'El huésped' and Domestic Violence," *Grafemas*, December 2007, http://people.wku.edu/inma.pertusa/encuentros/grafemas/diciembre_07/pennington.html.
71. Gómez-Barris, *The Extractive Zone*, xvi.
72. Svampa, *Neo-Extractivism in Latin America*, 16.
73. Veronica Gerber Bicecci, *La compañía* (Mexico DF: Almadia, 2020), no pagination. Recent history illustrates the continuation of this dynamic. Mexico's largest gold mine, Peñasquito, located in Mazapil, Zacatecas, and run by the multinational Goldcorp, has been the focus of many protests. Community members have protested Goldcorp's low wages as well as its contamination of the water supply. Yet the government has repeatedly sided with Goldcorp over community concerns. Public and private security forces intimidate protestors through repression and criminalization. These tactics seek to muddy the waters, smearing protestors as "delinquent groups" akin to organized crime, rather than concerned citizens. According to the Observatory of Mining Conflicts in Latin America (OCMAL), Mexico has the highest number of ongoing mining conflicts in Latin America: 55 as of June 2020, out of 277 in Latin America. Todd Miller, *Empire of Borders: The Expansion of the US Border around the World* (New York: Verso, 2019), 168–69.
74. Alfredo Valadez, "Nuevo Mercurio, una mina de contaminación," *Zacatecas Online*, August 24, 2010. https://zacatecasonline.com.mx/index.php/noticias/municipios/6932-mina-nuevo-mercurio-un-cementerio-toxico. Although

Mexican law prohibited the import of hazardous waste for storage or disposal (only permitting its recycling), Luis Vera-Morales observes that "waste finds its way in through the backdoor." Some waste that enters Mexico is managed by compliant recyclers, but more often importers and exporters alike are aware that the transaction is fraudulent. The offloading of waste by rich nations onto poorer ones is symptomatic of global environmental racism. NAFTA purported to address this issue, but simply garbed it in legal protection. From 1995 to1997, 161 illegal hazardous dump sites were found in eighteen different states in Mexico. This problem is far from solved. A 2008 report by the US Government Accountability Office found that the United States has done nothing to impede the illegal export of waste. Luis R. Vera-Morales, "Dumping in the International Backyard: Exportation of Hazardous Wastes to Mexico," *Tulane Environmental Law Journal* 7, no. 2 (1994): 355; "NAFTA's Impact on North American Hazardous Waste Imports and Exports," *Hazardous Waste Consultant* 19, no. 3 (2001): 2.16–2.19; Tania Volke Sepúlveda and Jan Antonio Velasco Trejo, *Tecnologías de remediación para suelos contaminados* (Mexico: Institutio Nacional de Ecología, 2002), 26.

75. Álvaro García Hernández, "El pueblo fantasma de Nuevo Mercurio y la minería insostenible," *La Jornada Zacatecas*, July 9, 2014, http://ljz.mx/2014/07/09/el-pueblo-fantasma-de-nuevo-mercurio-y-la-mineria-insostenible.

76. Alfredo Valadez Rodríguez, "Mina zacatecano, convertida en un cementerio tóxico," *La Jornada*, August 23, 2010, https://www.jornada.com.mx/2010/08/23/estados/031n1est.

77. Rogelio Costilla-Salazar et al., "Assessment of Polychlorinated Biphenyls and Mercury Levels in Soil and Biological Samples from San Felipe, Nuevo Mercurio, Zacatecas, Mexico," *Bulletin of Environmental Contamination and Toxicology* 86, no. 2 (2010), https://doi.org/10.1007/s00128-010-0165-z, 214.

78. Even if one leaves the zone of contamination, toxicity lives on in the body. "The nature of metal poisoning, accumulated over decades," Mel Chen writes from their own experience, "is that any and every organ, including my brain, can bear damage. Because symptoms can reflect the toxicity of any organ, they form a laundry list that includes cognition, proprioception, emotion, agitation, muscle strength, tunnel perception, joint pain, and nocturnality. . . . The transformation by a toxin and its companions can be so comprehensive that it renders their host somewhat unrecognizable." Mel Chen, *Animacies: Biopolitics, Racial Mattering, and Queer Affect* (Durham, NC: Duke University Press, 2012), 201, 205.

79. Gerber Bicecci uses ecohorror to diagnose the ills of capitalism in a way that disentangles horror and the disabled body, which is more problematically collapsed in Samanta Schweblin's *Distancia de rescate*, in which the narrator seems less able to see the horror of industry but is horrified by the disfigured bodies of the children born in toxin-saturated areas.

80. Michelle Murphy, "Alterlife and Decolonial Chemical Relations," *Cultural Anthropology* 32, no. 4 (2017), 498.

81. Gerber Bicecci, *La compañía*. Translations are my own.
82. Studnicki-Gizbert, "Exhausting the Sierra Madre," 20.
83. Víctor M. Toledo, *Ecocidio en México: La batalla final es por la vida* (Mexico DF: Grijalbo 2015), 49.
84. Eve Tuck, "Suspending Damage: A Letter to Communities," *Harvard Educational Review* 79, no. 3 (2009), 412.
85. Journalist Álvaro García Hernández noted some of these desires in a piece from 2014: "En el futuro inmediato, los habitantes de la citada comunidad aspiran a cosas imposibles como tener agua potable, contar con vías de comunicación adecuadas, un transporte público de calidad, seguridad pública y servicios de salud eficientes" (In the immediate future, the inhabitants of the aforementioned community aspire to imposible things like potable wáter, adequate means of communication, quality public transportation services, public safety, and efficient healthcare). Álvaro García Hernández, "El pueblo fantasma de Nuevo Mercurio y la minería insostenible," *La Jornada Zacatecas*, July 9, 2014, http://ljz.mx/2014/07/09/el-pueblo-fantasma-de-nuevo-mercurio-y-la-mineria-insostenible.
86. Gerber Bicecci, *La compañía*.
87. Gerber Bicecci changes the maid to the machine in her rewriting, which might seem to dissipate the importance of interclass political alliances in taking down the Company. But viewed through the lens of Benjamin's theory of disjunctive percolation, the sedimented subsurface of Dávila's original configuration of feminist interclass revenge continues to inform the story's revised meaning. As a character that helps take down the Company, the machine's presence suggests that technology is not inherently problematic; technologies can be put to work to serve communities rather than to serve the interests of extractivist capital.
88. Svampa, *Neo-Extractivism in Latin America*, 38.
89. Gerber Bicecci, *La compañía*.
90. Angela Ndalianis, *The Horror Sensorium: Media and the Senses* (Jefferson, NC: McFarland, 2012), 3.
91. Elizabeth Barrios, "This Is Not an Oil Novel: Obstacles to Reading Petronarratives in High-Energy Cultures," *Textual Practice* 35, no. 3 (2021), 369. Only recently has anti-extractivist Mexican art begun to take a self-reflexive turn. See Carolyn Fornoff, "Reflexive Extractivist Aesthetics," *FORMA* 2, no. 1 (2023): 37–69.
92. Although Dávila's short story is a story of feminist solidarity in the face of domestic abuse, it nonetheless centers the housewife's point of view. While the maid's child is the guest's first victim and the maid is essential in defeating the guest, the class hierarchy between housewife and maid remains untroubled at the story's close, even after the guest's defeat. Similar class dynamics are at play in Gerber Bicecci's rewriting. The prophetic account of the fear the reader will experience when extraction encroaches upon their normalcy attempts to bridge the distance between those who live in the extractive zone and the largely urban public of Gerber Bicecci's art.
93. Velázquez, "Entrevista Verónica Gerber Bicecci."

94. Graciela Speranza, *Lo que no vemos, lo que el arte ve* (Buenos Aires: Anagrama, 2022).
95. Benjamin, *The Arcades Project*, 392.
96. Barthes, *Image-Music-Text*, 148.
97. Rivera Garza, *The Restless Dead*, 65.

CHAPTER 2

1. In July 2020, a judge ruled that the pipeline could not pass through Amilcingo, since it (a Nahua community that largely opposed the project) was not provided with voting booths for the referendum. The judge found that this violated the right of informed consultation. The pipeline will likely be rerouted through nearby towns. Madeleine Wattenbarger, "The Legacy of Samir Flores, One Year Later," *NACLA*, February 18, 2020, https://nacla.org/news/2020/02/18/legacy-samir-flores-one-year-later.
2. Itxaro Arteta, "Huexca dice no a la consulta y a la termoeléctrica; gobierno presume éxito de participación," *Animal Político*, February 24, 2019, https://www.animalpolitico.com/2019/02/huexca-consulta-termoelectrica-rechazo-protestas.
3. Jaime Osorio, "Crisis estatales y violencia desnuda: La excepcionalidad mexicana," in *Violencia y crisis del estado: Estudio sobre éxico*, ed. Jaime Osorio (México DF: Universidad Autónoma Metropolitana, 2011), 57.
4. Jaime Osorio, *Reproducción del capital, estado y sistema mundial: Estudios desde la teoría marxista de la dependencia* (Bogotá: Universidad Nacional de Colombia, Facultad de Derecho, Ciencias Políticas y Sociales, 2017).
5. See the multimedia project in honor of Samir Flores Soberanes, "#SamirVive," accessed February 2, 2022, https://www.samirvive.art/home.
6. Idelber Avelar, *The Untimely Present: Postdictatorial Latin American Fiction and the Task of Mourning* (Durham, NC: Duke University Press, 1999), 2.
7. Avelar, *The Untimely Present*, 2.
8. My understanding of the hauntological is informed by Derrida, who thinks about it as an as-of-yet unrealized promise inherited from the past, as well as a responsibility to those who have come before and those who are yet to come. In this sense, the hauntological operates in a subjunctive key, always straining against what is with the specter of what could have been or could still be. Jacques Derrida, *Specters of Marx: The State of the Debt, the Work of Mourning, and the New International*, trans. Peggy Kamuf (New York: Routledge Classics, 2006).
9. Catherine Gallagher, "Undoing," in *Time and the Literary*, ed. Karen Newman, Jay Clayton, and Marianne Hirsch (New York: Routledge, 2002), 11–12.
10. Rocío González de Arce, "La naturaleza vista por el cine mexicano," *Ciencia* 69, no. 2 (2018), 24.
11. Rebecca Janzen, "*El Cambio/The Change* Joskowicz ([1971] 1975): Mexican Counterculture and the Futility of Protest in the 1970s," *Studies in Spanish & Latin American Cinemas* 18, no. 2 (January 2021): 159–75.
12. Jacqueline Bixler, "Mexico 68 and the Art(s) of Memory," in *The Long 1968: Revisions and New Perspectives*, ed. Daniel J. Sherman et al. (Bloomington: Indiana University Press, 2013).

13. Achille Mbembe, *Necropolitics*, trans. Steve Corcoran (Durham, NC: Duke University Press, 2019), 36.
14. Oswaldo Zavala, *Los cárteles no existen: Narcotráfico y cultura en México* (Barcelona: Malpaso, 2018), 23 (my translation).
15. Dawn Paley, *Drug War Capitalism* (Edinburgh: AK Press, 2015), 19. The militarization of these regions under the guise of the drug war, Paley explains, is motivated by "the acquisition of territory and resources, including increased control over social worlds and labor power."
16. Cristina Rivera Garza, *Grieving: Dispatches from a Wounded Country*, trans. Sarah Booker (New York: Feminist Press, 2020), 22.
17. Rivera Garza, *Grieving*, 4.
18. The surge in violence against land defenders has been linked by scholars to the expansion of extractivism, deregulation, and impunity, which allows such crimes to continue without fear of consequence. Globally, only 10 percent of land defender murders result in a conviction, compared with a 43 percent conviction rate in homicides. In Mexico, according to Global Witness, only one percent of reported crimes against land defenders are resolved. Nathalie Butt et al., "The Supply Chain of Violence," *Nature Sustainability* 2, no. 8 (2019): 742–47, doi:10.1038/s41893-019-0349-4, 743.
19. Svampa, *Neo-Extractivism in Latin America*, 7.
20. Tetreault, "New Extractivism in Mexico," 3. Tetreault notes that between 2000 and 2012, 31 percent of Mexican territory was involved in some sort of concession.
21. For a longer history of how Indigenous conceptualizations of territory have been delegitimized, see Jorge Quintana-Navarrete, "Reading Race in Rocks: Political Geology in Nineteenth-Century Mexico," *Journal of Latin American Cultural Studies* 30, no. 4 (2021): 525–43.
22. Nathalie Butt et al., "Supply Chain of Violence," 742.
23. Centro Mexicano de Derecho Ambiental, *Informe sobre la situación de las personas y comunidades defensoras de los derechos humanos ambientales en México, 2022* (Mexico: CEMDA, 2023), 9.
24. Global Witness, "Decade of Defiance," September 2022, https://www.global-witness.org/en/campaigns/environmental-activists/decade-defiance/#a-global-analysis-2021. Global Witness and CEMDA have different ways of tracking lethal and non-lethal attacks against land defenders, with CEMDA only identifying twenty-five killings in 2021. Perhaps this lower count can be attributed to CEMDA's more limited understanding of land defense, in contrast with Global Witness' intersectional approach to human rights defense and land defense. Regardless of these differences, both reports agree that violence against land defenders in Mexico is on the rise.
25. Centro Mexicano de Derecho Ambiental, *Informe . . . 2022*, 11.
26. Global Witness, "Decade of Defiance." Budget cuts explain the escalation in attacks against land defenders since 2020. Under the economic squeeze prompted by the COVID-19 pandemic, in 2020 the budget for the Comisión

Nacional de Áreas Naturales Protegidas was reduced by 75 percent, reducing already weak regulation of protected areas from illegal logging and mining. The legislature similarly did away with state funds to support land defenders, like the Fondo para la Protección de Personas Defensoras de Derechos Humanos y Periodistas and the Fondo Sectorial de Investigación Ambiental. CEMDA, *Informe sobre la situación de las personas defensoras de los derechos humanos ambientales en México, 2020* (Mexico: CEMDA, 2021), 13.

27. Rita Laura Segato, "A Manifesto in Four Themes," trans. Ramsey McGlazer, *Critical Times* 1, no. 1 (2018): 202.
28. In an unusual move, after confrontations with protestors, President Andrés Manuel López Obrador affirmed in April 2021 that his government would not allow the planned hydroelectric dam to proceed and that the state would shift its energy projects toward wind (an equally contentious industry in Oaxaca). Zósimo Camacho, "Canceladas, presas La Parota y Paso de la Reyna," *Contralínea*, 14 July 2021, https://www.contralinea.com.mx/archivo-revista/2021/04/15/canceladas-presas-la-parota-y-paso-de-la-reyna.
29. Rob Nixon, *Slow Violence and the Environmentalism of the Poor* (Cambridge, MA: Harvard University Press, 2013).
30. Rivera Garza, *Grieving*, 2.
31. Mariana Mora, "Desparición forzada, racismo institucional y pueblos indígenas en el caso Ayotzinapa, México," *LASA Forum* 48, no. 2 (2017), 29.
32. Impunity intersects with racism. One notable example is the acquittal of three men identified as involved in the 2018 forced disappearance of Nahua land defender Sergio Rivera Hernández, who exposed human rights violations during the construction of the Coyolapa-Atzatlán hydroelectric project in Sierra Negra, Puebla. The judge associated with the case was accused of making racist remarks to witnesses. "Juez concede libertad a los responsables identificados de la desaparición forzosa de Sergio Rivera," Frontline Defenders, September 15, 2020, https://www.frontlinedefenders.org/es/case/disappearance-sergio-rivera-hernandez#case-update-id-12392.
33. Yásnaya Elena A. Gil, "*Jëtsuk*: Nuestro ambientalismo se llama defensa del territorio," *El País*, April 3, 2021, https://elpais.com/mexico/2021-04-04/jetsuk-nuestro-ambientalismo-se-llama-defensa-del-territorio.html.
34. Organizations like the National Assembly of the Environmentally Affected (ANAA) are changing this landscape. ANAA is a national network of communities and environmentalists that formed in 2008 and provides a space for alliances between affected communities, academics, scientists, and environmentalists. Gabriela Méndez Cota, "Policing the Environmental Conjuncture: Structural Violence in Mexico and the National Assembly of the Environmentally Affected," *New Formations* 96 (2019), 69–88.
35. Toledo, Garrido, and Barrera-Bassols, "Struggle for Life," 133–47.
36. Henri Lefebvre, "Space and the State," in *State, Space, World*, trans. Gerald Moore, Neil Brenner, and Stuart Elden (Minneapolis: University of Minnesota Press, 2009), 224.

37. See the definition of *ayllu* in Marisol de la Cadena, *Earth Beings: Ecologies of Practice across Andean Worlds* (Durham, NC: Duke University Press, 2015), 101; or the concept of *ts'iib* in Paul M. Worley and Rita M. Palacios, *Unwriting Maya Literature: Ts'íib as Recorded Knowledge* (Tucson: University of Arizona Press, 2019), 11.
38. Floriberto Díaz, *Escrito: Comunalidad, energía viva del pensamiento mide. Ayuujktsënää'yën - ayuujkwënmää'ny - ayuujk mëk'äjtën*, ed. Sofía Robles Hernández and Rafael Cardoso Jiménez (México: Universidad Nacional Autónoma de México, 2007), 26.
39. Díaz, *Escrito*, 41.
40. Guillermo Bonfil Batalla, *México Profundo: Reclaiming a Civilization*, trans. Philip A. Dennis (Austin: University of Texas Press, 1996).
41. Quoted in Ana Matías Rendón, "Contra el despojo territorial," *Ojarasca* 264 (April 2019), http://ojarasca.jornada.com.mx/2019/04/12/contra-el-despojo-264-8847.html.
42. In the context of Amazonian literature, Amanda Smith finds cultural extractivism to be the structuring logic behind many literary texts written by mestizo *letrados* who mobilize Indigenous concepts yet bypass Indigenous interlocutors in their critiques of extractivist projects. This sort of "literary extractivism," Smith posits, is often misread as bringing greater visibility or awareness to Indigenous thought. In reality, this elision ignites yet another cycle of commodification, this time mining "cultural and spiritual resources" in order to circulate them "in literary and touristic markets." The effect for Smith is the "objectification and systematic selection of idealized Indigenous spiritual practices and ontologies by foreigners [which ameliorates] the anxieties of 'modern' living elsewhere." Amanda M. Smith, *Mapping the Amazon: Literary Geography after the Rubber Boom* (Liverpool, UK: Liverpool University Press, 2021), 145.
43. Ana Matías Rendón, "Contra el despojo territorial," *Ojarasca* 264 (April 2019): 3–6, http://ojarasca.jornada.com.mx/2019/04/12/contra-el-despojo-264-8847.html.
44. Mikeas Sánchez, "Jujtzyere' / Cuánto vale?" in *Jujtzye tä wäpä tzamapänh'ajä / Cómo ser un buen salvaje* (Guadalajara: Universidad de Guadalajara, 2019), 83–84. Published in translation: Wendy Call, trans., "What Is It Worth?," *Modern Poetry in Translation*, no. 3 (November 2019). https://modernpoetryintranslation.com/poem/what-is-it-worth.
45. Sánchez, "Jujtzyere' / Cuánto vale?"
46. "Muere trabajador en accidente en la mina de San José del Progreso," *La Minuta* (blog), EDUCA, August 20, 2019, https://www.educaoaxaca.org/muere-trabajador-en-accidente-en-la-mina-de-san-jose-del-progreso.
47. Backlash to Minera Cuzcatlán's sponsorship of Oaxaca FilmFest in the form of sustained protest and boycotts resulted in the cancellation of this sponsorship. Samantha Demby, "Mining Culture Wars Escalate in Oaxaca," NACLA, December 2, 2019, https://nacla.org/news/2019/12/01/mining-culture-wars-oaxaca-tourism.
48. Paloma Villanueva, "La mina que dividió a un pueblo," Oxfam México, May 10, 2018, https://www.oxfammexico.org/historias/la-mina-que-dividi%C3%B3-un-pueblo.

In tandem with overt repression, extractive companies use "softer" coercive measures and indirect counterinsurgency strategies that "mitigate conflict through sophisticated public relations efforts, [use] formal or informal funds to promote social fragmentation... and force [Indigenous communities] to either negotiate, accept the company's offer, or fight to stop the arriving projects." These intimidation tactics have spread throughout Oaxaca, as communal Indigenous lands are seized without consent by the Secretary of National Defense to build military bases near extractivist sites, or through forms of "green grabbing" that justify the expropriation of communal land to implement private, transnational wind energy projects. Alexander Dunlap affirms that these efforts have coalesced into a form of "sustainable violence" in which police-military repression campaigns aligned with renewable energy projects have become a permanent feature of the political landscape. Alexander Dunlap, "Wind Energy: Toward a 'Sustainable Violence' in Oaxaca," *NACLA Report on the Americas* 49, no. 4 (2017), 484, 488.

49. Rivera Garza, *Grieving*, 7.
50. Emmanuel Levinas, *Totality and Infinity: An Essay on Exteriority*, trans. Alphonso Lingis (Pittsburgh, PA: Duquesne University Press, 1969), 198.
51. Laura González-Flores, "What Is Present, What Is Visible: The Photo-Portraits of the 43 'Disappeared' Students of Ayotzinapa as Positive Social Agency," *Journal of Latin American Cultural Studies* 27, no. 4 (2018): 499.
52. Susan Sontag, "Posters: Advertisement, Art, Political Artifact, Commodity," in *The Art of Revolution: Castro's Cuba: 1959–1970*, ed. Dugald Stermer (New York: McGraw-Hill, 1970), vii–xxiii.
53. Mikeas Sánchez explains how the term "vandals" has been used by the mainstream media to describe protestors like the *normalistas rurales* in Ayozinapa, Mactumatzá, and Teteles. This delegitimizes their calls for education and social justice by vilifying them, painting them as the problematic social element, rather than the state that oppresses them. Mikeas Sánchez, "Me sumo a los que llaman 'vándalos' de las normales rurales," *Ojarasca* 290 (June 2021): 6–7, https://www.jornada.com.mx/2021/06/12/ojarasca290.pdf.
54. Sandra Walklate, *Imagining the Victim of Crime* (New York: Open University Press, McGraw-Hill, 2007), 28.
55. Henri Lefebvre, *The Production of Space*, trans. Donald Nicholson-Smith (Malden, MA: Blackwell, 1991), 170.
56. The reference to seeds is borrowed from the Zapatista National Army of Liberation (EZLN) phrase, "Quisieron enterrarnos, pero no sabían que éramos semillas" (They wanted to bury us, but they didn't know that we were seeds). The Zapatistas, an anticapitalist Indigenous social movement and political group that controls areas in Chiapas, themselves borrowed this phrasing from the Greek poet Dinos Christianopolous, who wrote in the 1970s, "You did anything to bury me. But you forgot I was a seed." Alternatively the phrase has been traced back to an epitaph written by Ernesto Cardenal for Adolfo Báez Bone, "Te mataron y no nos dijieron donde enterraron su cuerpo, pero desde

entonces todo el territorio nacional es tu sepulcro; o más bien; en cada palmo de territorio nacional en que no está tu cuerpo, tú resucitaste. Creyeron que te mataban con una orden de ¡fuego! Creyeron que te enterraban y lo que hacían era enterrar una semilla." (They killed you and they didn't tell us where they buried your body, but since then all the national territory is your grave; or more likely; in every measurement of the land where your body isn't, you were reborn. They thought that they killed you with the order to fire! They thought that they buried you when what they did was bury a seed.) Ever since, the phrase has been ubiquitous in activist work, invoked in Ayotzinapa protests by thousands on social media and in public protest as a performative materialization of the "we" that refuses to be silenced. It has been adopted generatively by artists, as in the eponymous 2014 work by trans performance artist Daniel Brittany Chávez, who inserted forty-three needles into his back. It has traveled beyond Mexico as a phrase used in transnational activism, by the Uruguayan socialist politician Daniel Martínez, the Black Lives Matter movement in the United States, and in the Transgender Memorial Garden in St. Louis. See Marcela A. Fuentes, *Performance Constellations: Networks of Protest and Activism in Latin America* (Ann Arbor: University of Michigan Press, 2019), 5; Daniel B Coleman and Rolando Vázquez, "Precedence, Trans* and the Decolonial," in *Tranimacies: Intimate Links between Animal and Trans* Studies*, ed. Eliza Steinbock, Marianna Szczygielska, and Anthony Clair Wagner (London: Routledge, 2021); Jill H. Casid, "Necrolandscaping," in *Natura: Environmental Aesthetics after Landscape*, ed. Jens Andermann, Lisa Blackmore, and Dayron Carrillo Morell (Zurich: Diaphanes, 2018), 240.

57. Jorge A Pérez Alfonso, "Borran mural de arte urbano en el centro histórico de Oaxaca," *La Jornada*, January 29, 2016, https://www.jornada.com.mx/2016/01/29/cultura/a03n1cul.

58. Marisol Rojas, "The Multidimensional and the Multiple in Contemporary Art: *Let's Sow Dreams, Let's Harvest Hope* (2015–2018) by Lapiztola Stencil," Marisol Rojas / portfolio, accessed August 2, 2021, http://marisolrojas.com/textos_2.html. Rojas discusses the mural's afterlife and resignification as part of the project Coachella Walls, an initiative led by Armando Lerma in Coachella's Historic Pueblo Viejo district. This time, Cariño's text was removed, and under the new title "American Woman," the image was reframed as a tribute to undocumented Chicana women. It was also subsequently repurposed to support the candidacy of Nahuatl leader Marichuy in the 2018 presidential elections.

59. Alice Driver, *More or Less Dead: Feminicide, Haunting, and the Ethics of Representation in Mexico* (Tucson: University of Arizona Press, 2015), 105.

60. Red Mexicana de Afectadas/os por la Minería, "Bety Cariño, nos arrebataron tu presencia y hasta hoy la impunidad impera," Facebook, April 27, 2021, https://www.facebook.com/REMAMX/photos/pcb.5476222835753242/5476222549086604.

61. Félix García, "Corrido a Bety Cariño," https://www.youtube.com/watch?v=MfswuzXy9I4; Alberta Cariño Trujillo, *Poemas, escritos y discursos de Bety*

Cariño (1976–2010) (Oaxaca: El Rebozo, 2013); Ruma Barbero, *Sembrando sueños, cosechando esperanzas: La historia de Bety Cariño, indígena feminista, activista, defensora de los derechos de la madre tierra y de los pueblos* (Pamplona: Mugarik Gabe, 2017); *Homenaje a Bety Cariño*, directed by Iván Castaneira, 2016, https://www.youtube.com/watch?v=xnHEE5GL6tU; *Bety Cariño, tejedora de esperanza*, CACTUS production, 2011, https://www.youtube.com/watch?v=DeZvO1drNxY&t=132s; *Murha Meksikossa*, directed by Simo Sipola (Helsinki: Yleisradio, 2016).

62. *The Formaldehyde Trip* was also performed live at the 2018 Pacific Standard Time Festival: Live Art LA/LA series *En Cuatro Patas* curated by Neo Bustamante and Sandra Ibarra. For analysis of this series, see Bernadine Hernández, "Living on All Fours: Latinx Performance and the Trans Human Turn in *Cuatro Patas*," *Transgender Studies Quarterly* 6, no. 2 (January 2019): 261.

63. Laura G. Gutiérrez, "*Resiliencia tlacuache* by Naomi Rincón Gallardo," *Terremoto*, May 11, 2020, https://terremoto.mx/en/online/resilencia-tlacuache.

64. Nicole Seymour, *Bad Environmentalism: Irony and Irreverence in the Ecological Age* (Minneapolis: University of Minnesota Press, 2018), 120–21.

65. Gómez-Barris, *The Extractive Zone*.

66. Tuck, "Suspending Damage," 417.

67. Alfredo López Austin, *Los mitos del tlacuache: Caminos de la mitología mesoamericana* (Mexico DF: Universidad Autónoma de México, 1996).

68. Jill Dolan, *Utopia in Performance: Finding Hope at the Theater* (Ann Arbor: University of Michigan Press, 2005), 7.

69. Naomi Rincón Gallardo in discussion with the author, June 2021, Oaxaca.

70. Walter Benjamin, *Reflections: Essays, Aphorisms, Autobiographical Writings*, trans. Edmund Jephcott (Boston, MA: Mariner Books, 2019), 189.

71. Anita Gonzalez, "Indigenous Acts: Black and Native Performances in Mexico," *Radical History Review*, no. 103 (2009): 135.

72. Naomi Rincón Gallardo, "*The Formaldehyde Trip*: A Mythical/Critical (Under)world-Making Dedicated to Bety Carino," *Critical Ethnic Studies* 4, no. 2 (2018): 39.

73. This quote is from lyrics sung by the axolotl in the subchapter "Enigma." Rincón Gallardo writes of the queer, racialized love between Alex and Axol: "They encounter each other within the logic and structure of racist practices, which arrange the world under a particular racial ordering within which Axol is supposed to respond to Alex's needs and commands. Yet their fleeting encounter is intimate, poignantly charging the surface of contact between the two of them with desires situated on the edge of the dominant orders of belonging and subjugation." Rincón Gallardo, "*The Formaldehyde Trip*," 47.

74. Rincón Gallardo, interview with the author, 39.

75. Alberta Cariño Trujillo, *La poesía es resistencia, es otra manera de seguir: Poemas, escritos y discursos de Bety Cariño (1976–2010)* (Oaxaca: El Rebozo, Palapa Editorial, 2013), 6.

76. Cecilia F. Klein, "Fighting with Femininity: Gender and War in Aztec Mexico," *Estudios de cultura náhuatl*, no. 24, (1994): 225–26.
77. Klein, "Fighting with Femininity," 227.
78. Klein, "Fighting with Femininity," 236.
79. Gloria E. Anzaldúa, *Light in the Dark / Luz en lo oscuro: Rewriting Identity, Spirituality, Reality*, ed. AnaLouise Keating (Durham, NC: Duke University Press, 2015).
80. Cherríe Moraga, *The Last Generation: Prose and Poetry* (Boston, MA: South End Press, 1993), 72.
81. Gutiérrez, "*Resilencia tlacuache*."
82. Saidiya Hartman, "Venus in Two Acts," *Small Axe: A Caribbean Journal of Criticism* 12, no. 2 (2008): 11.
83. Federico Navarrete, "El lugar de las siete cuevas," *Revista de la Universidad de México* 80 (February 2019): 86.
84. John Law, "What's Wrong with a One-World World?" *Distinction* 16, no. 1 (2015): 126–39.
85. Adam Lifshey, *Specters of Conquest: Indigenous Absence in Transatlantic Literatures* (New York: Fordham University Press, 2010), 125.
86. Arturo Escobar, *Designs for the Pluriverse: Independence, Autonomy, and the Making of Worlds* (Durham, NC: Duke University Press, 2017), 6.
87. Irmgard Emmelhainz, "Decolonial Love," *e-Flux*, no. 99 (April 2019), https://www.e-flux.com/journal/99/262398/decolonial-love.
88. Mbembe, *Necropolitics*, 37.

CHAPTER 3

1. Gerardo Ceballos, Paul R. Ehrlich, and Rodolfo Dirzo, "Biological Annihilation via the Ongoing Sixth Extinction Signaled by Vertebrate Population Losses and Declines," *PNAS* 114, no. 30 (2017): E6089–96.
2. Banco de México, *Resultados de los grupos focales 2015 de la Dirección General de Emisión* (Mexico: Oficina de Análisis y Estudios de Efectivo, 2015), https://www.banxico.org.mx/billetes-y-monedas/d/%7B816B50E3-CCA1-3571-0D2C-8AF750FA3CD6%7D.pdf.
3. Banco de México, *Resultados de la segunda ronda de Grupos Focales 2019 de la Dirección General de Emisión* (Mexico: Oficina de Análisis y Estudios de Efectivo, 2019), 11. https://www.banxico.org.mx/billetes-y-monedas/d/%7BCAAA9C74-C078-1610-017A-E915895E7038%7D.pdf.
4. Lala, Twitter, Jan 7, 2022, https://twitter.com/yoongillauren2/status/1479662502166372357.
5. Sam Schipani, "How to Save the Paradoxical Axolotl," *Smithsonian Magazine*, Jan 8, 2018, https://www.smithsonianmag.com/science-nature/saving-paradoxical-axolotl-180967734.
6. Isabel Zapata, "Miembro fantasma," in *Una ballena es un país* (Mexico: Almadía, 2019), 51.

7. Jean Baudrillard, *Simulations*, trans. Paul Foss, Paul Patton, and Philip Beitchman (New York: Semiotext(e), 1983), 11.
8. Zapata makes the case for the capaciousness of poetry in *Una ballena es un país*. The second poem in the collection, "En el estrecho de Puget" (In Puget Sound), narrates the story of Richard Russell, who in 2018 hijacked an airplane from the Seattle-Tacoma airport in order to commit suicide. Zapata homes in on a specific moment in the tape recordings of Russell's communication with air traffic control. The transcripts show Russell asking for the coordinates of Tahlequah, an orca whale who at that time was famously carrying the corpse of her dead calf in an unusual display of grief. In the concluding verse, Zapata imagines that Russell was able to glimpse the whale and her dead calf in the water, a brief moment of connection between species experiencing despair. She opens the concluding verse that conjures this speculative scene with the line: "Como el poema lo permite todo, imaginemos que" (Since a poem permits anything, let us imagine that). This striking line delineates poetry's capaciousness as a space in which everything, from transcripts to fiction, is permitted. Poetry is capacious enough to operate simultaneously in the realm of fact and fiction—"let us imagine that." Zapata, *Una ballena es un país*, 16.
9. David Farrier, *Anthropocene Poetics: Deep Time, Sacrifice Zones, and Extinction* (Minneapolis: University of Minnesota Press, 2019), 5.
10. Mario Aquilina, "Electronic Literature and the Poetics of Contiguity," *The Bloomsbury Handbook of Electronic Literature*, edited by Joseph Tabbi (London: Bloomsbury, 2017), 205.
11. Emmanuel Levinas, "Language and Proximity," in *Collected Philosophical Papers*, trans. Alphonso Lingus (Pittsburgh, PA: Duquesne University Press, 1998), 109–26.
12. Patrick Wolfe, "Settler Colonialism and the Elimination of the Native," *Journal of Genocide Research* 8, no. 4 (Dec. 2006): 388.
13. Jason W. Moore, "Ecology, Capital, and the Nature of Our Times: Accumulation and Crisis in the Capitalist World-Ecology," *Journal of World Systems Research* 17, no. 1 (2011), 114.
14. Serge Dedina, *Saving the Gray Whale: People, Politics, and Conservation in Baja California* (Tucson: University of Arizona Press, 2000), 59.
15. Ursula K. Heise, *Imagining Extinction: The Cultural Meanings of Endangered Species* (Chicago: University of Chicago Press, 2016), 68.
16. José Emilio Pacheco, "Ballenas," in *Islas a la deriva: Poemas, 1973–1975* (México DF: Era, 2006), 86.
17. José Emilio Pacheco, "Augurios" and "Zopilote," in *Nuevo álbum de zoología* (Mexico DF: Era, 2013), 45, 50–51. For more on animals and environmental crisis in Pacheco, see Randy Malamud, "José Emilio Pacheco: 'I saw a dying fish,'" in *Poetic Animals and Animal Souls* (New York: Palgrave, 2003), 77–92; Scott Devries, *Creature Discomfort* (Leiden: Koninklijke Brill, 2016), 158–66; Micah McKay, "'Pasto sin fin del basurero': Trash and Disposal in the Poetry of José Emilio Pacheco," *Latin American Literary Review* 47, no. 93 (2020): 49–58.

18. Homero Aridjis and Betty Ferber, *Noticias de la tierra* (Mexico: Debate, 2012), 22.
19. Dedina, *Saving the Gray Whale*, 61.
20. Aridjis's collection *El ojo de la ballena* (The eye of the whale), published in 2001, celebrated the defeat of the contentious salt factory that would have been installed by Mitsubishi at Laguna San Ignacio, potentially interfering with gray whale breeding grounds. For more on extinction in Aridjis, see Scott DeVries, *Creature Discomfort*, 151–58; Heise, *Imagining Extinction*, 46–48; and Adam Spires, "Homero Aridjis and Mexico's Eco-Critical Dystopia," in *Blast, Corrupt, Dismantle, Erase: Contemporary North American Dystopian Literature*, ed. Brett Josef Grubisic, Gisele M. Baxter, and Tara Lee (Waterloo, Canada: Wilfred Laurier University Press, 2014), 339–54.
21. Emily Apter, *Against World Literature: On the Politics of Untranslatability* (New York: Verso, 2013), 353.
22. While nostalgia is often reactionary—a yearning for the past that reinforces the status quo—Aridjis's recollection of a time when monarch butterflies were less threatened is more aligned with what Jennifer Ladino calls "counternostalgia": nostalgia founded on the need for change. Jennifer Ladino, *Reclaiming Nostalgia: Longing for Nature in American Literature* (Charlottesville: University of Virginia Press, 2012), 15.
23. Anna Kathryn Kendrick, "Homero Aridjis, Public Pedagogy and an Educational Poetics of Environmentalism," *Bulletin of Hispanic Studies* 100, no. 1 (2023): 97–116. See also Juan Carlos Galeano and Laura Parcés, "Introduction: Eco-poetry and Eco-Art from Latin America," *Global South* 16, no. 1 (2023): 9–12.
24. "AMLO defiende construcción del Tren Maya: No afectará ecosistemas del sureste," *Forbes México*, June 12, 2020, https://www.forbes.com.mx/politica-amlo-defiende-construccion-del-tren-maya-no-afectara-ecosistemas-del-sureste.
25. For years these groups have petitioned the government to address regional drought, which has been exacerbated over the last thirty years by climate change and deforestation, affecting flora and fauna, as well as hunting and subsistence farming. The plan to transport 3 million tourists a year to an underserved, vulnerable region is so obscene that the Zapatistas (Ejército Zapatista de Liberación Nacional) have called it "an act of war." Before his death, Zapotec artist Francisco Toledo condemned the train as an "ecological disaster," and renowned activist Gustavo Esteva explained that it will "systematically dismantle the way of life" of local Indigenous peoples. Part of the problem, ecologist Julia Carabias signals, is that the project will establish two new cities in Calakmul and Bacalar of fifty thousand people each to serve the tourist influx. Opponents fear that the strain on water caused by a tourist boom will disproportionately affect Indigenous communities, and even lead to their "extinction."

    To this, AMLO responded that his critics "lacked information." The patronizing dismissal ironically held a grain of truth: the UN Commission on Human Rights concluded that Indigenous communities were indeed left out of the megaproject's decision-making process. In a vote held in December 2019, only 2.8 percent of the region's Indigenous population participated. The UN also observed that discussions about the Maya Train implicitly promised underserved communities with services like water and medical care, but did not write these promises into legislation. As Nahua activist María de Jesús Patricio Martínez (Marichuy) has argued, the government's lack of consideration for what Indigenous groups want shows that the state approaches its Indigenous peoples in extractive terms, valuing them for their "folkloric" color, rather than actually engaging with their desires.

Jorge A. Pérez Alfonso and Mónica Mateos-Vega, "Construir el Tren Maya 'va a ser un desastre,' alerta Francisco Toledo," *La Jornada*, Feb. 6, 2019, https://www.jornada.com.mx/2019/02/06/cultura/a03n1cul; Gustavo Esteva, "El atropello redentor," *La Jornada*, June 29, 2020, https://www.jornada.com.mx/2020/06/29/opinion/018a2pol; "EZLN rechaza Tren Maya por que destruirá su territorio; 'les falta información,' responde AMLO," Animal Político, Jan. 2, 2020 https://www.animalpolitico.com/2020/01/ezln-rechaza-tren-maya-falta-informacion-responde-amlo; Blanca A. Camargo and Mario Vázquez-Maguirre, "Humanism, Dignity and Indigenous Justice: The Mayan Train Megaproject, Mexico," *Journal of Sustainable Tourism* 29, no. 2-3 (2021), 8; Jorge Morla, "Marichuy: 'De los indígenas, al Gobierno mexicano solo le interesa el folclore,'" *El país*, 12 Oct. 2019, https://elpais.com/internacional/2019/10/12/actualidad/1570904742_707998.html.

26. Impacts on wildlife range from the introduction of nonnative species to the increase of illicit extraction. Train vibrations could collapse regional cave systems, which constitute Mexico's largest freshwater reserve. These aquifers maintain hydrological equilibrium, filtering water for mangroves, which in turn protect coral reefs from hurricanes.
27. Leonardo Domínguez, "Mayan Train puts more than 2,000 jaguars at risk," *El Universal*, Jan. 1, 2019, https://www.eluniversal.com.mx/english/mayan-train-puts-more-2000-jaguars-risk.
28. "AMLO defiende construcción del Tren Maya."
29. Ignacio Sánchez Prado, "Lengua precaria: La poesía mexicana en crisis," *América sin nombre* 23 (2018): 50.
30. Alejandro Higashi, "Hitos provisionales en el perfil de una generación: Poetas mexicanos nacidos entre 1975 y 1985," *Literatura mexicana* 25, no. 2 (2014), 57. It should be noted that Isabel Zapata's *Una ballena es un país* was reprinted by Almadía in 2021, demonstrating the bestiary's continued appeal.
31. Higashi, "Hitos provisionales," 63.
32. Pedro Uc Be, *Resistencia del territorio maya frente al despojo* (Mexico: Centro de Estudios para el Cambio en el Campo Mexicano, 2021).
33. Irmgard Emmelhainz, *Toxic Loves, Impossible Futures: Feminist Living as Resistance* (Nashville, TN: Vanderbilt University Press, 2022), 156–57.
34. Gisela Heffes, "Estéticas del Antropoceno: Tres preguntas con Isabel Zapata," Hablemos escritoras, Dec. 27, 2021, https://www.hablemosescritoras.com/posts/719.

35. Thom Van Dooren, *Flight Ways: Life and Loss at the Edge of Extinction* (New York: Columbia University Press, 2014), 3–4.
36. Zapata, *Una ballena*, 51.
37. Charles Hoge, "The Dodo in the Long Eighteenth Century: An Exploration of the Gray Ghost outside of the English Sentimental Eye," *University of Toronto Quarterly* 83, no. 3 (2014): 701.
38. Hoge, "The Dodo in the Long Eighteenth Century," 702.
39. Prior to the arrival of the Dutch the island was uninhabited; it had been visited by Arab traders in the Middle Ages and the Portuguese in the early sixteenth century, but neither settled there, nor did they leave written records of the bird. Van Dooren, *Flight Ways*, 2.
40. Karen Villeda, *Dodo* (Mexico: Fondo Editorial Tierra Adentro, 2013), 42.
41. Villeda, *Dodo*, 11.
42. In the context of global shipping at the time, labor was a commodity to be bought and sold, discursively inscribed by how the "maritime proletariat…were called 'hands.'" Marcus Rediker, *Outlaws of the Atlantic: Sailors, Pirates, and Motley Crews in the Age of Sail* (Boston, MA: Beacon, 2014), 5.
43. Karl Marx, *Capital: A Critique of Political Economy*, vol. 1, translated by Samuel Moore and Edward Aveling (Mineola, NY: Dover, 2011), 85.
44. Robert Kurz, "Domination without a Subject" (1993), quoted in Daniel Cunha, "The Geology of the Ruling Class?" *Anthropocene Review* 2 no. 3 (2015): 262–66.
45. Villeda, *Dodo*, 42. Italics in original.
46. Alejandro Higashi, "Karen Villeda habla sobre *Dodo*," *Ancila: Crítica de Poesía Mexicana Contemporánea*, 1 June 2014. See also Julio E. Ruiz Monroy, "*Dodo* la supervivencia de la poesía extinta," *Luvina* 76 (2014).
47. Villeda, *Dodo*, 54.
48. Heise, *Imagining Extinction*, 61.
49. Xitlálitl Rodríguez Mendoza, *Jaws (Tiburón)* (Guadalajara: Mantis, 2015), 13.
50. Rodríguez Mendoza, *Jaws*, 13.
51. Rodríguez Mendoza, *Jaws*, 21.
52. Shelley C. Clarke, Shelton J. Harley, Simon D. Hoyle, and Joel S. Rice, "Population Trends in Pacific Oceanic Sharks and the Utility of Regulations on Shark Finning," *Conservation Biology* 27, no. 1 (2013): 197–209.
53. Shelley Clarke, E. J. Milner-Gulland, and Trond Bjorndal, "Social, Economic, and Regulatory Drivers of the Shark Fin Trade," *Marine Resource Economics* 22, no. 3 (2007): 305–27.
54. Francisco Arreguín-Sánchez and Enrique Arcos-Huitrón, "La pesca en México: Estado de la explotación y uso de los ecosistemas" *Hidrobiológica* 21, no. 3 (2011): 431–62.
55. Oscar Sosa-Nishizaki et al., "Conclusions: Do We Eat Them or Watch Them, or Both? Challenges for Conservation of Sharks in Mexico and the NEP," *Advances in Marine Biology* 85 (2020): 97.
56. Sosa-Nishizaki et al., "Conclusions," 97.

57. Rubén D. Arvizu, "Tiburones en peligro de extinción por pesca ilegal, ligado al tráfico humano," Animal político, May 1, 2019. https://www.animalpolitico.com/blog-invitado/tiburones-en-peligro-de-extincion-por-pesca-ilegal-ligado-al-trafico-humano.
58. Rodríguez Mendoza, *Jaws*, 20.
59. Rodríguez Mendoza, *Jaws*, 22.
60. Rodríguez Mendoza, *Jaws*, 24.
61. Guerrero, *El sueño de toda célula*, 20.
62. Guerrero, *El sueño de toda célula*, 97.
63. Guerrero, *El sueño de toda célula*, 16.
64. Guerrero, *El sueño de toda célula*, 42.
65. Pamela Maciel Cabañas, "El lobo mexicano," in "Extinción," *Revista de la Universidad de México*, November 2017, 50. My translation. Mónica Nepote has also argued that one of the reasons that wolves have been hunted is their ontological relationship with territory: "piensan con su cuerpo, y en este llevan incrustado el territorio" (think with their bodies, and in their bodies their territory is embedded). Mónica Nepote, "Humanos y no humanos: Lenguajes emitidos, pensamientos escuchados," *Las repúblicas de lo salvaje* (blog), Aug. 18, 2022, https://lasrepublicasdelosalvaje.blog/blog/humanos-y-no-humanos-lenguajes-emitidos-pensamientos-escuchados.
66. Guerrero, *El sueño de toda célula*, 36.
67. Guerrero, *El sueño de toda célula*, 77.
68. Guerrero, *El sueño de toda célula*, 67.
69. US Fish and Wildlife Service, "Service Finalizes Changes to Mexican Wolf Experimental Population Rule in Arizona and New Mexico," press release, January 12, 2015.
70. Guerrero, *El sueño de toda célula*, 74, 16.
71. Gilles Deleuze and Félix Guattari, *A Thousand Plateaus: Capitalism and Schizophrenia*, trans. Brian Massumi (Minneapolis: University of Minnesota Press, 2003), 34.
72. Guerrero, *El sueño de toda célula*, 78.
73. Guerrero, *El sueño de toda célula*, 102.
74. Guerrero, *El sueño de toda célula*, 66.
75. Zapata, *Una ballena es un país*, 51.
76. For an overview of these debates, see Robert R. McKay, "Representation," in *Critical Terms for Animal Studies*, ed. Lori Gruen, 307–19 (Chicago: University of Chicago Press, 2018).
77. Héctor Hoyos, *Things with a History: Transcultural Materialism and the Literatures of Extraction in Contemporary Latin America* (New York: Columbia University Press, 2019), 38.
78. Jacques Derrida, *Acts of Literature*, ed. Derek Attridge (New York: Routledge, 1992), 83.
79. Derrida, *Acts of Literature*, 83.

80. Eve Kosofsky Sedgwick, *Touching Feeling: Affect, Pedagogy, Performativity* (Durham, NC: Duke University Press, 2003), 8.

CHAPTER 4

1. Barbie Zelizer, *About to Die*, 13.
2. Roland Barthes, *Camera Lucida: Reflections on Photography*, trans. Richard Howard (New York: Hill and Wang, 1981), 45.
3. Zelizer, *About to Die*, 13.
4. Zelizer, *About to Die*, 23.
5. Eduardo Zepeda, Timothy A. Wise, and Kevin P. Gallagher, "Rethinking Trade Policy for Development: Lessons from Mexico under NAFTA," *Carnegie Policy Outlook*, Dec. 2009, https://carnegieendowment.org/2009/12/07/rethinking-trade-policy-for-development-lessons-from-mexico-under-nafta-pub-24271.
6. Mark Weisbrot, Stephan Lefebvre, and Joseph Sammut, "Did NAFTA Help Mexico? An Assessment after 20 Years," *CEPR Reports and Issue Briefs*, Center for Economic and Policy Research, Feb. 2014, https://www.cepr.net/documents/nafta-20-years-2014-02.pdf.
7. Guillermo N. Murray-Tortarolo, "Seven Decades of Climate Change across Mexico," *Atmósfera* 34, no. 2 (2021): 217–26; Laura Gottesdiener, "Dams, Taps Running Dry in Northern Mexico amid Historic Water Shortages," Reuters, June 20, 2022, https://www.reuters.com/world/americas/dams-taps-running-dry-northern-mexico-amid-historic-water-shortages-2022-06-20.
8. INEGI, "Población rural y urbana," Cuéntame de Mexico, 2020, accessed February 2, 2022. http://cuentame.inegi.org.mx/poblacion/rur_urb.aspx?tema=P. Urban ecodocs that take place in Mexico City have focused primarily on two issues: earthquakes and water scarcity. See Mark Anderson, "The Grounds of Crisis and the Geopolitics of Depth: Mexico City in the Anthropocene," in *Ecological Crisis and Cultural Representation in Latin America: Ecocritical Perspectives on Art, Film, and Literature*, ed. Mark Anderson and Zélia M. Bora (Lanham, MD: Lexington, 2016), 99–124; and Oscar A. Pérez, "Local Landscapes, Global Conversations: The Case of Three Environmental Documentary Films from the Hispanic World," *Hispanic Issues On Line*, no. 24 (2019): 66–79.
9. Shuaizhang Feng, Alan B. Krueger, and Michael Oppenheimer, "Linkages among Climate Change, Crop Yields and Mexico-US Cross-Border Migration," *Proceedings of the National Academy of Sciences* 107, no. 32 (2010): 14, 257–62, https://doi.org/10.1073/pnas.1002632107. This study argues that out-migration from Mexico between 1995 and 2005 (the great Mexican emigration) has been attributed to labor changes and NAFTA, but that one overlooked factor is drought in northern Mexico after 1994, which was as severe as the great Mexican drought of the 1950s. In 2021, 85 percent of the country was experiencing extreme drought, comparable only to the 2011 drought that affected 95 percent of the country. "Widespread Drought in Mexico," NASA Earth Observatory (NASA), accessed February 2, 2022, https://earthobservatory.nasa.gov/images/148270/widespread-drought-in-mexico.

10. Kanta Kumari Rigaud et al., *Groundswell: Preparing for Internal Climate Migration* (Washington, DC: World Bank, 2018), 73–74.
11. Abrahm Lustgarten, "Where Will Everyone Go?" ProPublica, July 23, 2020, https://features.propublica.org/climate-migration/model-how-climate-refugees-move-across-continents.
12. For a robust critique of how the climate migrant has become a problematic trope of "liberal humanitarian concern" that "[evacuates] the agency of migrants, [removes] focus from the proximate political and economic situations in their home contexts, and [narrates] their actions against the ghostly backdrop of the coming uninhabitability of their places of origin," see Neel Ahuja, *Planetary Specters: Race, Migration, and Climate Change in the Twenty-First Century* (Chapel Hill: University of North Carolina Press, 2021). For more on migration cinema, see Christina L. Sisk, *Mexico Nation in Transit: Contemporary Representations of Mexican Migration to the United States* (Tucson: University of Arizona Press, 2011).
13. See Darren Thomas, Terry Mitchell, and Courtney Arseneau, "Re-evaluating Resilience: From Individual Vulnerabilities to the Strength of Cultures and Collectivities Among Indigenous Communities," *Resilience* 4, no. 2 (2016): 116–29.
14. Mark Vardy and Mick Smith, "Resilience," *Environmental Humanities* 9, no. 1 (2017): 177.
15. Rigaud et al., *Groundswell*.
16. Vardy and Smith, "Resilience."
17. In the twentieth century, films with environmental content were dramatically lower: González de Arce estimates a total of five films in the 1970s, seven in the '80s, and twenty-one in the '90s. Rocío Betzabeé González de Arce Arzave, "El viaje del cine mexicano de ficción hacia la conciencia ecológica: Imaginarios de la naturaleza, ecoutopías y ética ambiental en la pantalla," (MA thesis, Universidad Iberoamericana, 2019), 140.
18. Other domestic festivals also welcome films about environmental concerns, such as the Festival de Cine en el Campo (founded in 2008) and the Encuentros Hispanoamericanos de Cine y Video Documental Independiente Contra el Silencio de Todas las Voces (founded in 2000), which has a competition category dedicated to sustainability. Lauro Zavala notes that seven festivals dedicated to the Mexican documentary have been created since 2000, which promote the genre's creation and consumption. Lauro Zavala, "El nuevo documental mexicano y las fronteras de la representación," *Toma Uno* 1 (2012): 30.
19. Environmental film festivals build audiences by bringing together local and international communities around a shared set of concerns. Kay Armatage has suggested that festivalgoer interest in ecodocs demonstrates that interest in environmental issues is not sufficiently satisfied by the mainstream media. In Latin America, mainstream coverage of local environmental struggles is highly politicized. Jorge Marcone explains that the media is overwhelmingly critical

of resistance to national development projects and reinforces state narratives that deem land defenders obstacles to growth. Within this context, the ecodoc is a genre through which to tell overlooked stories, contest official narratives, and cultivate external support. Kay Armatage, "Planet in Focus: Environmental Film Festivals," in *Screening Nature: Cinema beyond the Human*, ed. Anat Pick and Guinevere Narraway (New York: Berghahn Books, 2013), 264; and Jorge Marcone, "Filming the Emergence of Popular Environmentalism in Latin America: Postcolonialism and Buen Vivir," in *Global Ecologies and the Environmental Humanities: Postcolonial Approaches*, ed. Elizabeth DeLoughrey, Jill Did, and Anthony Carrigan (New York: Routledge, 2015), 214.

20. Jorge Caballero, "*Cuates de Australia* no aborda la sequía, sino un pueblo que espera la lluvia," *La Jornada*, Jan. 31, 2013, https://www.jornada.com.mx/2013/01/31/espectaculos/a08n1esp.

21. Caballero, "*Cuates de Australia*."

22. Jens Andermann, "The Politics of Landscape," in *A Companion to Latin American Cinema*, ed. Maria M. Delgado, Stephen M. Hart, and Randal Johnson (Hoboken, NJ: John Wiley and Sons, 2017), 134.

23. *Cuates de Australia*, directed by Everardo González (Ciénaga, 2013).

24. Andermann, "The Politics of Landscape," 135, 138.

25. Alberto Anaya Adalid and Everardo González, *Cuates de Australia* (Mexico DF: Elefanta, 2014).

26. Tamara L. Falicov, "'Cine en Construcción' / 'Films in Progress': How Spanish and Latin American Film-Makers Negotiate the Construction of a Globalized Art-House Aesthetic," *Transnational Cinemas* 4, no. 2 (2013): 253–71, doi:10.1386/trac.4.2.253_1, 261.

27. Miriam Ross, "The Film Festival as Producer: Latin American Films and Rotterdam's Hubert Bals Fund," *Screen* 52, no. 2 (2011): 263, doi:10.1093/screen/hjr014, 263.

28. Ignacio Manuel Sánchez Prado, "The Politics-Commodity: The Rise of Mexican Commercial Documentary in the Neoliberal Era," in *Latin American Documentary Film in the New Millennium*, ed. María Guadalupe Arenillas and Michael J. Lazzarra (New York: Palgrave, 2016), 103.

29. Roberto Forns-Broggi, "Los retos del ecocine en nuestras Américas: Rastreos del buen vivir en tierra sublevada," *Revista de Crítica Literaria Latinoamericana* 40, no. 79 (2014): 323.

30. Marcone, "Filming the Emergence," 209.

31. On extractive forms of cultural consumption, see Ramón Grosfoguel, "Del 'extractivismo económico' al 'extractivismo ontológico': Una forma destructiva de conocer, ser y estar en el mundo," *Tabula Rasa* 24 (2016): 123–43.

32. John Patrick Leary, "Keywords for the Age of Austerity 19: Resilience," Keywords: The New Language of Capitalism, June 23, 2015, https://keywordsforcapitalism.com/2015/06/23/keywords-for-the-age-of-austerity-19-resilience.

33. For a robust critique of the humanitarian genre, see Pooja Rangan, *Immediations: The Humanitarian Impulse in Documentary* (Durham, NC: Duke University Press, 2017).
34. Paul Sbrizzi, "A Conversation with Everardo González (*Drought*)," *Hammer to Nail*, August 24, 2012, https://www.hammertonail.com/interviews/a-conversation-with-everardo-gonzalez-drought.
35. Sbrizzi, "A Conversation."
36. This is evidenced by director Laura Herrero Garvín's shift from the more traditionally constructed *Son duros los días sin nada* (*The Days with Nothing Are Rough*, co-directed with Laura Salas Sánchez, 2012), a testimonial portrait of women who led efforts to repair their communities in the wake of natural disasters, to the more sensorial *El Remolino*, which tells a similar story of survival in environmentally precarious conditions, but this time by privileging ambiance and contemplative visuals paired with reflexive techniques like participatory film-within-film footage.
37. A recent documentary about the extinction of the cardenche musical genre that also interweaves environmental issues affecting northern Mexico is *A morir a los desiertos*, directed by Marta Ferrer (Mexico: Cuadernos de Cineb, 2017).
38. Jesús Adolfo Soto Curiel, *Recordar en presente: Cine documental y memoria en México* (Mexicali: Universidad Autónoma de Baja California, 2017), 48.
39. Scott MacDonald, *American Ethnographic Film and Personal Documentary: The Cambridge Turn* (Berkeley: University of California Press, 2013), 315.
40. Jens Andermann, *Tierras en trance* (Santiago, Chile: Metales Pesados, 2018).
41. Jonathan Romney, "In Search of Lost Time," *Sight and Sound* 20, no. 2 (2010): 44.
42. Sayak Valencia, *Gore Capitalism*, trans. John Pluecker (Los Angeles: Semiotext(e), 2018).
43. Jorge Caballero, "*Cuates de Australia* no aborda la sequía, sino un pueblo que espera la lluvia," *La Jornada*, January 31, 2013, https://www.jornada.com.mx/2013/01/31/espectaculos/a08n1esp.
44. Tomás Crowder-Taraborrelli, "Migration, Regional Traditions, and the Intricacy of Documentary Representation in *Cuates de Australia* and *La chica del sur*," *Latin American Perspectives* 41, no. 1 (2013): 172–76. Germán Martínez Martínez has elaborated a similar critique, arguing that *Cuates* perpetuates the stereotype that rural Mexico is saturated in superstition. Germán Martínez Martínez, "El cine de Everardo González y las posibilidades del documental en México," in *Reflexiones sobre cine mexicano contemporáneo: Documental*, ed. Claudia Curiel de Icaza and Abel Muñoz Hénonin (Mexico DF: Cineteca Nacional, 2014), 61. Also see Martha Torres Méndez, *Desert Landscapes: Violence and Enduring Subjectivities in Contemporary Mexican Literary and Visual Cultures* (PhD diss., UC Irvine, 2020). For more on masculinity in Mexican cinema, see Samantha Ordóñez, *Mexico Unmanned: The Cultural Politics of Masculinity in Mexican Cinema* (Albany, NY: SUNY Press, 2022).
45. Sbrizzi, "A Conversation."

46. Sbrizzi, "A Conversation." Francisco Javier Ramírez-Miranda argues that González's inclusion of moments where subjects interact with the camera indicates that the ranchers had a great deal of sway in dictating how their community was represented. But we could equally posit these are merely included to acknowledge the agential presence of the filmmaker shaping the narrative on screen. Francisco Javier Ramírez-Miranda, "Aproximaciones a lo real en el cine de Everardo González," *Comunicación y Medios* 36 (2017): 33–42.
47. Caballero, "Cuates de Australia."
48. Flavio Lehner et al., "Projected Drought Risk in 1.5°C and 2°C Warmer Climates," *Geophysical Research Letters* 44, no. 14 (2017): 7419–28, https://doi.org/10.1002/2017gl074117.
49. "2011, el peor año de sequía en México," Animal Político, June 7, 2011, https://www.animalpolitico.com/2011/06/2011-el-peor-ano-de-sequia-en-mexico.
50. Carlos Escalante-Sandoval and Pedro Nuñez-Garcia, "Meteorological Drought Features in Northern and Northwestern Parts of Mexico under Different Climate Change Scenarios," *Journal of Arid Land* 9, no. 1 (2016): 65–75.
51. J. María Mendoza-Hernández et al., "Proyecciones climáticas para el estado de Coahuila usando el modelo precis bajo dos escenarios de emisiones," *Agrociencia* 47 (2013): 523–37.
52. Jennifer Fay, *Inhospitable World: Cinema in the Time of the Anthropocene* (Oxford, UK: Oxford University Press, 2018), 17.
53. Carlos Monsiváis, "Yes, nor Do the Dead Speak, Unfortunately: Juan Rulfo," in *Mexican Postcards*, trans. John Kraniauskus (London: Verso, 1997), 59.
54. Ageeth Sluis, *Deco Body, Deco City: Female Spectacle and Modernity in Mexico City, 1900–1939* (Lincoln: University of Nebraska Press, 2016), 102.
55. Niamh Thornton, *Revolution and Rebellion in Mexican Film* (London: Bloomsbury, 2013), 115.
56. Sluis, *Deco Body, Deco City*, 18.
57. Monsiváis, "Yes, nor Do the Dead Speak," 60–61.
58. Emily Hind, "*Provincia* in Recent Mexican Cinema, 1989–2004," *Discourse* 26, no. 1–2 (Winter and Spring 2004), 26.
59. Ignacio Sánchez Prado, *Screening Neoliberalism: Transforming Mexican Cinema, 1988–2012* (Nashville, TN: Vanderbilt University Press, 2014), 54.
60. Caballero, "Cuates de Australia."
61. Brian L. Price, "Heterotemporal *Mise-en-scene* in the Films of Luis Estrada." *Arizona Journal of Hispanic Cultural Studies* 16 (2012): 259–74.
62. Víctor Martínez Ranero, review of *Los reyes del pueblo que no existe*, Time Out, Oct. 20, 2015, https://www.timeout.com/es/cine/los-reyes-del-pueblo-que-no-existe.
63. Val Plumwood, "Shadow Places and the Politics of Dwelling," *Australian Humanities Review* 44 (2008): 139–50.
64. Sibely Cañedo-Cázarez and Juan Manuel Mendoza-Guerrero, "Desplazamiento forzado y empoderamiento femenino: El caso de la presa Picachos en el sur de Sinaloa, México," *El Ágora U.S.B.* 17, no. 2 (2017): 371.

65. Betzabé García, dir., *Unsilenced*, New York Times Op-Docs, 10:02, March 22, 2016, https://www.nytimes.com/2016/03/22/opinion/unsilenced.html.
66. Lucy Bollington, "Landscapes of *desapropiación*: Necropolitics and Hydropoetics in Recent Mexican Documentary Film," *Journal of Romance Studies* 21, no. 2 (2021): 263–91.
67. Betzabé García, dir., *Los reyes del pueblo que no existe*, Venado Films and Ruta 66 Cine, 2015, https://vimeo.com/122781305.
68. Lilia Adriana Pérez Limón, "Documenting Precarity and Other Ghostly Remains: Passivity as Political Practice in Betzabé García's Documentary *Los reyes del pueblo que no existe*," *Studies in Latin American Popular Culture* 36 (2018): 96.
69. Sofía Viramontes, "Betzabé García y el devenir en la reconstrucción," *Gatopardo*, March 31, 2017, https://gatopardo.com/arte-y-cultura/betzabe-garcia-reyes-pueblo-no-existe.
70. Tamara L. Falicov, *Latin American Film Industries* (London: BFI, 2019), 51.
71. Betzabé García, dir., *Venecia, Sinaloa*, 2011, 6:42, https://vimeo.com/44362796.
72. Other Latin American films have also invoked Venice, like Emiliano Mazza de Luca's 2016 documentary *Nueva Venecia*, about a marshy submerged village in Colombia where soccer is played on stilts.
73. Erick Estrada, "FICM 2015: La angustia y la resistencia," Cinegarage, October 31, 2015, https://www.cinegarage.com/37100-ficm-2015-06; Rafael Paz, "'Los reyes del pueblo que no existe' y las visiones del apocalipsis," Butaca Ancha, November 30, 2015, http://butacaancha.com/los-reyes-del-pueblo-que-no-existe-y-las-visiones-del-apocalipsis.
74. Scott A. Kulp and Benjamin H. Strauss, "New Elevation Data Triple Estimates of Global Vulnerability to Sea-Level Rise and Coastal Flooding," *Nature Communications* 10 (2019).
75. Temitope D. Timothy Oyedotun, Arturo Ruiz-Luna, and Alma G. Navarro-Hernández, "Contemporary Shoreline Changes and Consequences at a Tropical Coastal Domain," *Geology, Ecology, and Landscapes* 2, no. 2 (2018): 104–14.
76. Pedro Ortega, "Betzabé García: *Los reyes del pueblo que no existe*," FilmArte, 2015, https://web.archive.org/web/20170508163611/http://www.filmarte.net/Entrevistas/betzabe-garcia-los-reyes-del-pueblo-que-no-existe.
77. Óscar Tinoco, "'Para mí el arte siempre va a tener que ver con la política': Betzabé García," Crash, May 16, 2016, http://www.crash.mx/encuadre/para-mi-el-arte-arte-siempre-va-a-tener-que-ver-con-la-politica-betzabe-garcia.
78. Falicov, *Latin American Film Industries*, 49.
79. Bill Nichols, *Representing Reality: Issues and Concepts in Documentary* (Bloomington: Indiana University Press, 1991), 38.
80. Laura Herrero Garvín, dir., *El Remolino* (IMCINE, 2016).
81. Donna Haraway, *Staying with the Trouble: Making Kin in the Chthulucene* (Durham, NC: Duke University Press, 2016), 132.
82. See Alyshia Gálvez, *Eating NAFTA: Trade, Food Policies, and the Destruction of Mexico* (Berkeley: University of California Press, 2018).

83. Allyse Knox-Russell, "Futurity without Optimism: Detaching from Anthropocentrism and Grieving Our Fathers in *Beasts of the Southern Wild*," in *Affective Ecocriticism: Emotion, Embodiment, Environment*, ed. Kyle Bladow and Jennifer Ladino (Lincoln: University of Nebraska Press, 2018), 213–32.
84. See Oswaldo Zavala, *Los cárteles no existen: Narcotráfico y cultura en México* (Barcelona: Malpaso, 2018).
85. Brad Evans and Julian Reid, "Dangerously Exposed: The Life and Death of the Resilient Subject," *Resilience: International Policies, Practices, and Discourses* 1, no. 2 (2013): 95.

CHAPTER 5

1. *El tema*, Episode 1, "Agua," directed by Santiago Maza, aired April 12, 2021, https://lacorrientedelgolfo.net/proyecto/el-tema/agua.
2. Ana Rosas Mantecón, *Ir al cine: Antropología de los públicos, la ciudad y las pantallas* (Mexico: Universidad Autónoma Metropolitana, 2017), 235.
3. CEDECINE, *Anuario Estadístico 2020* (Mexico: IMCINE, 2021), 88.
4. Juan Llamas-Rodriguez, "A Global Cinematic Experience: Cinépolis, Film Exhibition, and Luxury Branding," *JCMS* 58, no. 3 (2019): 49–71.
5. Winona LaDuke and Deborah Cowen, "Beyond Wiindigo Infrastructure," *South Atlantic Quarterly* 119, no. 2 (2020), 264.
6. Nadia Bozak, *The Cinematic Footprint: Lights, Camera, Natural Resources* (New Brunswick, NJ: Rutgers University Press, 2011).
7. This chapter is an effort to explore a different methodological approach to ecocinema than that advanced in my coedited volume: Carolyn Fornoff and Gisela Heffes, eds., *Pushing Past the Human in Latin American Cinema* (Albany, NY: SUNY Press, 2021).
8. Stephen Rust and Salma Monani, "Introduction: Cuts to Dissolves—Defining and Situating Ecocinema Studies," in *Ecocinema Theory and Practice*, ed. Stephen Rust, Salma Monani, and Sean Cubitt (New York: Routledge, 2013), 3.
9. Jussi Parikka, "New Materialism as Media Theory: Medianatures and Dirty Matter," *Communication and Critical/Cultural Studies* 9, no. 1 (2012): 95–100.
10. Fernando Coronil, *The Magical State: Nature, Money, and Modernity in Venezuela* (Chicago: University of Chicago Press, 1997); Santiago Acosta, "We Are Like Oil: An Ecology of the Venezuelan Culture Boom: 1973-1983" (PhD diss., Columbia University, 2020); Sean Nesselrode Moncada, *Refined Material: Petroculture and Modernity in Venezuela* (Oakland, CA: University of California Press, 2023).
11. Martina Broner, "Rethinking Format in the Amazon: Ecology and *El abrazo de la serpiente*," *JCMS* 61, no. 1 (2021): 25.
12. Sebastián Figueroa, "Apuntes sobre cine y extractivismo," *laFuga* 26 (2022): 3. https://www.lafuga.cl/apuntes-sobre-cine-y-extractivismo/1100.
13. Cajetan Iheka, *African Ecomedia: Network Forms, Planetary Politics* (Durham, NC: Duke University Press, 2021), 3-4.

14. *El tema*, episode 4, "Energía," directed by Santiago Maza, aired May 3, 2021, 11:23, https://lacorrientedelgolfo.net/proyecto/el-tema/energia.
15. Carolyn Fornoff, "Mexican Cinema as Petrocinema." *Studies in Spanish and Latin American Cinemas* 18, no. 3 (2021): 377–87.
16. José Alberto Hernández Ibarzábal and David Bonilla, "Examining Mexico's Energy Policy under the 4T," *Extractive Industries and Society* 7, no. 2 (2020): 669–75, 672.
17. Quoted in Vanessa Freije, *Citizens of Scandal: Journalism, Secrecy, and the Politics of Reckoning in Mexico* (Durham, NC: Duke University Press, 2020), 80.
18. Sánchez Prado, *Screening Neoliberalism*, 6.
19. Sánchez Prado, *Screening Neoliberalism*, 6.
20. Rosas Mantecón, *Ir al cine*, 243, my translation.
21. CEDECINE, *Anuario Estadístico 2020*, 138.
22. INEGI, "Comunicado de prensa," news release no. 352/21, June 22, 2021.
23. Fernando García-Mora and Jorge Mora-Rivera, "Exploring the Impacts of Internet Access on Poverty: A Regional Analysis of Rural Mexico," *New Media and Society* 25, no. 1 (Jan. 2023): 26–49.
24. Rosas Mantecón, *Ir al cine*, 19.
25. Laura U. Marks and Radek Przedpelski, "Bandwidth Imperialism and Small-File Media," Post45, April 13, 2021, https://post45.org/2021/04/bandwidth-imperialism-and-small-file-media.
26. Lotfi Belkhir and Ahmed Elmeligi, "Assessing ICT Global Emissions Footprint: Trends to 2040 and Recommendations," *Journal of Cleaner Production* 177 (2018): 448–63.
27. Marks and Przedpelski, "Bandwidth Imperialism."
28. Federico Demaria et al., "What Is Degrowth?: From an Activist Slogan to a Social Movement," *Environmental Values* 22, no. 2 (2013): 191–215. For a discussion of degrowth in Spain, see Luis I. Prádanos, *Postgrowth Imaginaries: New Ecologies and Counterhegemonic Culture in Post-2008 Spain* (Liverpool: Liverpool University Press, 2018), 64.
29. María Guadalupe Arenillas and Michael J. Lazzara, introduction to *Latin American Documentary Film in the New Millennium,* ed. María Guadalupe Arenillas and Michael J. Lazzara (New York: Palgrave, 2016), 1.
30. Claudia E. Saldaña-Durán et al., "Environmental Pollution of E-waste: Generation, Collection, Legislation, and Recycling Practices in Mexico," in *Handbook of Electronic Waste Management: International Best Practices and Case Studies*, ed. Majeti Narasimha Vara Prasad, Meththika Vithanage, Anwesha Borthakur (Oxford, UK: Butterworth-Heinemann, 2020), 422.
31. Verónica Gago, *Neoliberalism from Below: Popular Pragmatics and Baroque Economies*, translated by Liz Mason-Deese (Durham, NC: Duke University Press, 2017), 171.
32. Another alternative exhibition infrastructure initiative in Mexico is Ambulante, a nonprofit led by Gael García Bernal and Diego Luna. Ambulante focuses on

bringing Latin American documentary to viewers throughout Mexico, with the goal of fostering "a self-sustaining exhibition circuit that will continue to form film-viewing publics throughout the country year-round" (https://www.ambulante.org/iniciativas/presenta). They provide training to film programmers about how to transform community halls into screening spaces, and how to guide conversations with the public after screenings.

33. Freya Schiwy, *The Open Invitation: Activist Video, Mexico, and the Politics of Affect* (Pittsburgh. PA: University of Pittsburgh Press, 2019), 5.
34. Erick Ramírez, "Disruptores: Cine Móvil ToTo, la pantalla nómada," *El Sol de México*, March 12, 2020.
35. The cinema-by-pedaling model was pioneered in Latin America by the company Ecocinema in Uruguay (founded in 2010), with help from the Dutch foundation Solar Cinema (founded in 2006). Uruguay is also home to the Festival Internacional de Cine a Pedal (International Festival for Cinema by Pedal), a three-day film festival powered by twenty-two bikes that began in 2013 in conjunction with Cinemateca Uruguaya.
36. Tamara L. Falicov, "Mobile Cinemas in Cuba: The Forms and Ideology of Traveling Exhibitions," *Public* 40 (2012): 104.
37. Cine Móvil ToTo (website), accessed February 24, 2022, https://cinemoviltoto.mx/Index, my translation.
38. Hidalgo Neira, "Cine Móvil ToTo llevará cine, de manera sustentable, más allá de las grades ciudades," *Reporte Indigo*, August 18, 2021, https://www.reporteindigo.com/piensa/cine-movil-toto-llevara-cine-de-manera-sustentable-mas-alla-de-las-grandes-ciudades.
39. "Quiénes somos," Cine Móvil ToTo (website), accessed February 24, 2022, https://cinemoviltoto.mx/Quienes-Somos, my translation.
40. One of the world's largest lithium deposits has been discovered in Sonora. Lithium is also known as "white gold" or "the new oil." All current concessions for Sonora lithium mining were obtained by China's Ganfeng Lithium, and exempted by Andrés Manuel López Obrador from recent efforts to nationalize lithium mining. Ann Deslandes, "Mexico's Lithium and the Global Race to Lock in 'White Gold,'" *Al Jazeera*, December 21, 2021, https://www.aljazeera.com/economy/2021/12/21/mexicos-lithium-and-the-global-race-to-lock-in-white.
41. Echeverría, *Modernity and "Whiteness,"* 18.
42. Mónica Santillán Vera and Angel de la Vega Navarro, "Do the Rich Pollute More?: Mexican Household Consumption by Income Level and CO_2 Emissions," *International Journal of Energy Sector Management* 13, no. 3 (2019): 703.
43. Unlike bikes, solar panels are expensive to acquire and service. Solar panels also require the mining of lithium, an industry that Thea Riofrancos describes as "green extractivism," in which "human rights and ecosystems [are subordinated] to endless extraction in the name of 'solving' climate change." But this is not the only way that lithium could be managed. The future of lithium mining could go one of two ways, Riofranco observes. It could continue to

be implemented in a way that replicates current extractivist dynamics and energy consumption patterns oriented around "privatized and segregated suburban affluence," or it could be "a historic opportunity to dismantle [this] lifestyle and build something better in its place . . . reorganized to prioritize climate safety, socio-economic equality, Indigenous rights, and the integrity of habitats." In April 2022, Andrés Manuel López Obrador's administration nationalized lithium, offering a glimmer of hope for a more just management and equitable distribution of the metal. Mexico's nationalization of lithium presents an opportunity to delink state extractivism from accumulation and orient it instead around technology's potential to serve the commons. Thea Riofrancos, "What Green Costs," *Logic Magazine*, December 7, 2019. https://logicmag.io/nature/what-green-costs.

44. "Quienes somos," Aquí cine (website), accessed February 24, 2022, http://aquicineoaxaca.blogspot.com/p/quienes-somos.html.
45. Fernanda Río, *Manual para exhibicionistas: Cómo montar un cine* (Mexico DF: IMCINE, 2015), https://issuu.com/dennisehernandez/docs/manual_final01/1.
46. "Toolkit," Solar World Cinema, accessed September 15, 2022, http://www.solar-cinema.org/toolkit.
47. albert, *A Screen New Deal: A Route Map to Sustainable Film Production* (London: Albert, BFI, Arup, 2020), 6. https://wearealbert.org/wp-content/uploads/2021/03/Screen-New-Deal-Report-1.pdf.
48. albert, *Annual Review: May 2020–April 2021* (London: albert, 2021), 4. https://2021.wearealbert.org/albert_Annual_Review_2020%2021.pdf.
49. Carolyn Buchanan, "Carbon Footprint of Movie Production Location Choice: The Real Cost," (MA thesis, Harvard Extension School, 2016), 21.
50. Peter Flanigan, "The Environmental Cost of Filmmaking," *UCLA Entertainment Law Review* 10, no. 1 (2002): 69–95.
51. Ben Goldsmith and Tom O'Regan, *The Film Studio: Film Production in the Global Economy* (Lanham, MD: Rowman and Littlefield, 2005), 92.
52. Goldsmith and O'Regan, *The Film Studio*, 93.
53. Sophia McClennen, *Globalization and Latin American Cinema: Toward a New Critical Paradigm* (London: Palgrave Macmillan, 2018), 454.
54. Hunter Vaughan, *Hollywood's Dirtiest Secret: The Hidden Environmental Costs of the Movies* (New York: Columbia University Press, 2019), 1.
55. Goldsmith and O'Regan, *The Film Studio*, 79.
56. Buchanan, "Carbon Footprint," 28.
57. Ana Mónica Rodríguez, "*Bosque de niebla* muestra 'otra forma de relaciones humanas, de vivir en unidad,'" *La Jornada*, February 15, 2018, https://www.jornada.com.mx/2018/02/15/espectaculos/a09n1esp.
58. Mónica Álvarez Franco, "Ignite Session de Mónica Álvarez Franco (México)—Directora de 'Bosque de niebla,'" Los Cabos International Film Festival, YouTube, November 25, 2021, 8:55, https://www.youtube.com/watch?v=lwdMl49yWQI.

59. Christian Sandvig, "The Internet as the Anti-Television: Distribution Infrastructure as Culture and Power," in *Signal Traffic: Critical Studies of Media Infrastructures*, ed. Lisa Parks and Nicole Starosielski (Champaign: University of Illinois Press, 2015), 226.
60. Janet Walker and Nicole Starosielski, "Introduction: Sustainable Media," in *Sustainable Media: Critical Approaches to Media and Environment*, ed. Nicole Starosielski and Janet Walker (New York: Routledge, 2016), 1–20.

CONCLUSION

1. Enrique Leff, *Ecología y capital: Racionalidad ambiental, democracia participativa y desarrollo sustentable*, 6th ed. (México: Siglo XXI, 2005), 68.
2. Echeverría, *Modernity and "Whiteness,"* 73.
3. Echeverría, *Modernity and "Whiteness,"* 73.
4. Jens Andermann, *Tierras en trance: Arte y naturaleza después del paisaje* (Santiago: Metales Pesados, 2018).
5. Rachel Price, *Planet/Cuba: Art, Culture, and the Future of the Island* (New York: Verso, 2015).
6. Gómez-Barris, *The Extractive Zone*, xiv.
7. Rogers, *Mourning El Dorado*; Saramago, *Fictional Environments*, 13.
8. Guerrero, *El sueño de toda célula*, 78.
9. Guerrero, *El sueño de toda célula*, 77.
10. Elio Henríquez, "Piden revocar permiso de agua a FEMSA en Chiapas," *La Jornada*, June 25, 2020, https://www.jornada.com.mx/2020/06/25/politica/017n1pol.
11. Raúl Pacheco-Vega, "Agua embotellada en México: De la privatización del suministro a la mercantilización de los recursos hídricos," *Espiral* 22, no. 63 (2015): 221–63.
12. "FEMSA Biennial Will Hold Its 14th Edition in Michoacán," FEMSA, press release, January 14, 2020, https://www.femsa.com/en/press-room/press-release/the-femsa-biennial-will-hold-its-14th-edition-in-michoacan.
13. "FEMSA Foundation," Nature Conservancy, accessed February 26, 2022, https://www.nature.org/en-us/about-us/who-we-are/how-we-work/working-with-companies/companies-investing-in-nature1/femsa.
14. Jasbir K. Puar and Andrew Ross, "Decolonising the Museum," *Al Jazeera*, July 21, 2021, https://www.aljazeera.com/opinions/2021/7/21/decolonising-the.
15. Author conversation with Naomi Rincón Gallardo, summer 2021.
16. "Verónica Gerber: *La máquina distópica*," Seminario Investigación Poéticas de lo Inquietante, January 25, 2021, https://www.youtube.com/watch?v=tZtjCLTodTo&t=5845s. My translation.
17. "Verónica Gerber: *La máquina distópica*."
18. Alexis Shotwell, *Against Purity: Living Ethically in Compromised Times* (Minneapolis: University of Minnesota Press, 2016), 8–9.
19. Michael Rothberg, *The Implicated Subject: Beyond Victims and Perpetrators* (Redwood City, CA: Stanford University Press, 2019), 19.

20. Dení Noya, "Minera Cuzcatlán y Oaxaca FilmFest, crónica de un boicot anunciado," Avispa, October 12, 2019, https://avispa.org/minera-cuzcatlan-y-oaxaca-filmfest-cronica-de-un-boicot-anunciado.
21. Marán, "Piel territorio."

Bibliography

Acosta, Santiago. "We Are Like Oil: An Ecology of the Venezuelan Culture Boom: 1973–1983." PhD diss., Columbia University, 2020.
Aguilar Gil, Yásnaya Elena. "Jëtsuk: Nuestro ambientalismo se llama defensa del territorio." *El País*, April 3, 2021. https://elpais.com/mexico/2021-04-04/jetsuk-nuestro-ambientalismo-se-llama-defensa-del-territorio.html.
Ahuja, Neel. *Planetary Specters: Race, Migration, and Climate Change in the Twenty-First Century*. Chapel Hill: University of North Carolina Press, 2021.
albert. *Annual Review: May 2020–April 2021*. London: albert, 2021. https://2021.wearealbert.org/albert_Annual_Review_2020%2021.pdf.
———. *A Screen New Deal: A Route Map to Sustainable Film Production*. London: Albert, BFI, Arup, 2020. https://wearealbert.org/wp-content/uploads/2021/03/Screen-New-Deal-Report-1.pdf.
Álvarez Franco, Mónica. "Ignite Session de Mónica Álvarez Franco (México)—Directora de 'Bosque de niebla.'" Los Cabos International Film Festival, YouTube, November 25, 2021, 8:55. https://www.youtube.com/watch?v=lwdMl49yWQI.
Amatto, Alejandra. "Transculturar el debate. Los desafíos de la crítica literaria latinoamericana actual en dos escritoras: Mariana Enriquez y Liliana Colanzi." *Valenciana* 13, no. 26 (2020): 207–30.
"AMLO defiende construcción del Tren Maya: No afectará ecosistemas del sureste." *Forbes México*, June 12, 2020. https://www.forbes.com.mx/politica-amlo-defiende-construccion-del-tren-maya-no-afectara-ecosistemas-del-sureste.
Anaya Adalid, Alberto, and Everardo González. *Cuates de Australia*. Mexico DF: Elefanta, 2014.
Andermann, Jens. "The Politics of Landscape." In *A Companion to Latin American Cinema*, edited by Maria M. Delgado, Stephen M. Hart, and Randal Johnson, 133–49. Hoboken, NJ: John Wiley and Sons, 2017.
———. *Tierras en trance: Arte y naturaleza después del paisaje*. Santiago: Metales Pesados, 2018.

Anderson, Mark D. "The Grounds of Crisis and the Geopolitics of Depth: Mexico City in the Anthropocene." In *Ecological Crisis and Cultural Representation in Latin America: Ecocritical Perspectives on Art, Film, and Literature*, edited by Mark Anderson and Zélia M. Bora, 99–124. Lanham, MD: Lexington, 2016.

———. "Was the Mexican Revolution a Revolt of Nature?: Agustín Yáñez's Ecological Perspective." *Revista de Estudios Hispánicos* 40 (2006): 447–67.

Animal Político. "2011, el peor año de sequía en México." Animal Político, June 7, 2011. https://www.animalpolitico.com/2011/06/2011-el-peor-ano-de-sequia-en-mexico.

"Anthropocene in Chile: Toward a New Pact of Coexistence." *Environmental Humanities* 11, no. 2 (2019): 467–76.

Anzaldúa, Gloria E. *Light in the Dark / Luz en lo oscuro: Rewriting Identity, Spirituality, Reality*. Edited by AnaLouise Keating. Durham, NC: Duke University Press, 2015.

Apter, Emily S. *Against World Literature: On the Politics of Untranslatability*. London: Verso, 2013.

Arenillas, María Guadalupe, and Michael J. Lazzara. Introduction to *Latin American Documentary Film in the New Millennium*, edited by María Guadalupe Arenillas and Michael J. Lazzara, 1–20. New York: Palgrave, 2016.

Aridjis, Homero, and Betty Ferber. *Noticias de la tierra*. Mexico DF: Debate, 2012.

Aridjis, Homero, and Fernando C. Césarman. *Artistas e intelectuales sobre el ecocidio urbano*. Mexico DF: Consejo de la Crónica de la Ciudad de México, 1989.

Aridjis, Homero. *El ojo de la ballena: Poemas, 1999–2001*. Mexico DF: Fondo de Cultura Económica, 2014.

Armatage, Kay. "Planet in Focus: Environmental Film Festivals." In *Screening Nature: Cinema beyond the Human*, edited by Anat Pick and Guinevere Narraway, 257–74. New York: Berghahn Books, 2013.

Arreguín-Sánchez, Francisco, and Enrique Arcos-Huitrón. "La pesca en México: Estado de la explotación y uso de los ecosistemas." *Hidrobiológica* 21, no. 3 (September/December 2011): 431–62.

Arvizu, Rubén D. "Tiburones en peligro de extinción por pesca ilegal, ligado al tráfico humano." Animal Político, May 1, 2019. https://www.animalpolitico.com/blog-invitado/tiburones-en-peligro-de-extincion-por-pesca-ilegal-ligado-al-trafico-humano.

Barbas-Rhoden, Laura. *Ecological Imaginations in Latin American Fiction*. Gainesville: University of Florida Press, 2011.

Barbero, Ruma. *Sembrando sueños, cosechando esperanzas: La historia de Bety Cariño, indígena feminista, activista, defensora de los derechos de la madre tierra y de los pueblos*. Pamplona: Mugarik Gabe, 2017.

Barthes, Roland. *Camera Lucida: Reflections on Photography*. Translated by Richard Howard. New York: Hill and Wang, 1981.

———. *Image-Music-Text*. Translated by Stephen Heath. New York: Hill and Wang, 1977.

———. *The Preparation of the Novel: Lecture Courses and Seminars at the College de France (1978–1979 and 1979–1980)*. Translated by Kate Briggs. New York: Columbia University Press, 2011.

Barrios, Elizabeth. "This Is Not an Oil Novel: Obstacles to Reading Petronarratives in High-Energy Cultures." *Textual Practice* 35, no. 3 (2021): 363–78.

Baudrillard, Jean. *Simulations*. Translated by Paul Foss, Paul Patton, and Philip Beitchman. New York: Semiotext(e), 1983.

Beck, Humberto, Carlos Bravo Regidor, and Patrick Iber. "Year One of AMLO's Mexico." *Dissent*, Winter 2020. https://www.dissentmagazine.org/article/year-one-of-amlos-mexico.

Beckman, Ericka. *Capital Fictions: The Literature of Latin America's Export Age*. Minneapolis: University of Minnesota Press, 2013.

Bendeck, Geoff. "In the Absence of Words: An Interview with Verónica Gerber Bicecci." Words without Borders, February 21, 2018. https://www.wordswithoutborders.org/dispatches/article/in-the-absence-of-words-an-interview-with-veronica-gerber-bicecci.

Belkhir, Lotfi, and Ahmed Elmeligi. "Assessing ICT Global Emissions Footprint: Trends to 2040 and Recommendations." *Journal of Cleaner Production* 177 (2018): 448–63.

Bello, Andrés. *Gramática de la lengua castellana destinada al uso de los americanos: Obras completas. Tomo Cuatro*. Madrid: EDAF, 1984.

Benjamin, Walter. *The Arcades Project*. Translated by Howard Eiland and Kevin McLaughlin. Cambridge: Belknap Press of Harvard University Press, 1999.

———. *Reflections: Essays, Aphorisms, Autobiographical Writings*. Translated by Edmund Jephcott. Boston: Mariner Books, 2019.

Bixler, Jacqueline. "Mexico 68 and the Art(s) of Memory." In *The Long 1968: Revisions and New Perspectives*, edited by Daniel J. Sherman, Ruud Van Dijk, Jasmine Alinder, and A. Aneesh, 169–216. Bloomington: Indiana University Press, 2013.

Bohn, Willard. "The Visual Trajectory of Jose Juan Tablada." *Hispanic Review* 69, no. 2 (2001): 191–208. https://doi.org/10.2307/3247038.

Bollington, Lucy. "Landscapes of *desapropiación*: Necropolitics and Hydropoetics in Recent Mexican Documentary Film." *Journal of Romance Studies* 21, no. 2 (2021): 263–91.

Bournot, Estefanía. "Abrir las heridas. Gerber, Meruane y Mendieta: Geoescrituras de un planeta enfermo." *Revista Letral* 25 (2021): 54–73.

Boyer, Dominic. *Energopolitics: Wind and Power in the Anthropocene*. Durham, NC: Duke University Press, 2019.

Bozak, Nadia. *The Cinematic Footprint: Lights, Camera, Natural Resources*. New Brunswick, NJ: Rutgers Unviersity Press, 2011.

Broner, Martina. "Rethinking Format in the Amazon: Ecology and *El abrazo de la serpiente*." *JCMS: Journal of Cinema and Media Studies* 61, no. 1 (2021): 7–26.

Buchanan, Carolyn. "Carbon Footprint of Movie Production Location Choice: The Real Cost," MA thesis, Harvard Extension School (2016).

Buck-Morss, Susan. *The Dialectics of Seeing: Walter Benjamin and the Arcades Project*. Cambridge, MA: MIT Press, 1989.

Bugarin, Inder. "Natural Disasters: The High Cost of Climate Change." *El Universal*, October 11, 2018.

Burke, Marshall, Felipe González, Patrick Baylis, Sam Heft-Neal, Ceren Baysan, Sanjay Basu, and Solomon Hsiang. "Higher Temperatures Increase Suicide Rates in the United States and Mexico." *Nature Climate Change* 8 (2018): 723–29.

Butt, John, and Carmen Benjamin. *A New Reference Grammar of Modern Spanish*. New York: Routledge, 2011.

Caballero, Jorge. "*Cuates de Australia* no aborda la sequía, sino un pueblo que espera la lluvia." *La Jornada*, January 31, 2013. https://www.jornada.com.mx/2013/01/31/espectaculos/a08n1esp.

Camacho, Zósimo. "Canceladas, presas La Parota y Paso de la Reyna." *Contralínea*, April 15, 2021. https://www.contralinea.com.mx/archivo-revista/2021/04/15/canceladas-presas-la-parota-y-paso-de-la-reyna.

Camargo, Blanca A., and Mario Vázquez-Maguirre. "Humanism, Dignity and Indigenous Justice: The Mayan Train Megaproject, Mexico." *Journal of Sustainable Tourism* 29, no. 2-3 (2021): 372–91.

Campisi, Nicolás. "Documentary Mines: Archives of Ecohorror in the Anthropocene." In *Post-Global Aesthetics*, edited by Benjamin Loy and Gesine Müller, 131–48. Berlin: De Gruyter, 2023.

———. "Ruinas contemporáneas: Ficciones del eco-horror en América Latina." *Afuera*, June 4, 2020. https://alafuera.art.blog/2020/06/04/ruinas-contemporaneas-ficciones-del-eco-horror-en-america-latina.

Cañedo-Cázarez, Sibely, and Juan Manuel Mendoza-Guerrero. "Desplazamiento forzado y empoderamiento femenino: El caso de la presa Picachos en el sur de Sinaloa, México." *El Ágora* 17, no. 2 (2017): 371.

Cariño Trujillo, Alberta. *La poesía es resistencia, es otra manera de seguir: Poemas, escritos y discursos de Bety Cariño (1976–2010)*. Oaxaca: El Rebozo, Palapa Editorial, 2013.

Carle, Jill. "Climate Change Seen as Top Global Threat." Pew Research Center, July 14, 2015. https://www.pewresearch.org/global/2015/07/14/climate-change-seen-as-top-global-threat.

Carmona, Gisella L. *Ecologicon: Ecoliteratura en Alfonso Reyes*. Nuevo León: Universidad Autónoma de Nuevo Leon, 2012.

Casid, Jill H. "Necrolandscaping." In *Natura: Environmental Aesthetics after Landscape*, edited by Jens Andermann, Lisa Blackmore, and Dayron Carrillo Morell, 237–64. Zurich: Diaphanes, 2018.

Ceballos, Gerardo, Paul R. Ehrlich, and Rodolfo Dirzo, "Biological Annihilation via the Ongoing Sixth Extinction Signaled by Vertebrate Population Losses and Declines." *PNAS* 114, no. 30 (2017): E6089–E6096.

CEDECINE. *Anuario Estadístico 2020*. Mexico DF: IMCINE, 2021.

Centro Mexicano de Derecho Ambiental, *Informe sobre la situación de las defensoras de los derechos humanos ambientales en México, 2020*. Mexico: CEMDA, 2021.

———. *Informe sobre la situación de las personas y comunidades defensoras de los derechos humanos ambientales en México, 2022*. Mexico: CEMDA, 2023. https://www.cemda.org.mx/publicaciones-y-estudios-del-cemda/informe-sobre-la-situacion-de-las-personas-y-comunidades-defensoras-de-los-derechos-humanos-ambientales-en-mexico-2022.

Chen, Mel. *Animacies: Biopolitics, Racial Mattering, and Queer Affect*. Durham, NC: Duke University Press, 2012.

Cine Móvil ToTo (website), accessed February 24, 2022. https://cinemoviltoto.mx/Index.

———. "Quiénes somos," accessed February 24, 2022. https://cinemoviltoto.mx/Quienes-Somos, my translation.

Clarke, Shelley C., E. J. Milner-Gulland, and Trond Bjørndal. "Social, Economic, and Regulatory Drivers of the Shark Fin Trade." *Marine Resource Economics* 22, no. 3 (2007): 305–27.

Clarke, Shelley C., Shelton J. Harley, Simon D. Hoyle, and Joel S. Rice. "Population Trends in Pacific Oceanic Sharks and the Utility of Regulations on Shark Finning." *Conservation Biology* 27, no. 1 (February 2013): 197–209.

Cohen, José, and Lorenzo Hagerman, dirs. *H2Omx*. Mexico DF: Cactus Film & Video, 2014.

Coleman, Daniel B., and Rolando Vázquez. "Precedence, Trans* and the Decolonial." In *Tranimacies: Intimate Links between Animal and Trans* Studies*, edited by Eliza Steinbock, Marianna Szczygielska, and Anthony Clair Wagner, 38–44. London: Routledge, 2021.

Conn, Robert T. *The Politics of Philology: Alfonso Reyes and the Invention of the Latin American Literary Tradition*. Lewisburg, PA: Bucknell University Press, 2002.

"Conoce estos espectaculares alebrijes en el Zoológico de Chapultepec." Wipy.tv, March 30, 2018. https://wipy.tv/alebrijes-en-el-zoologico-de-chapultepec.

Contreras-Moreno, Fernando, and Yudith Torres-Ventura. "El cambio climático y los ungulados silvestres." *CICY* 10 (July 2018): 144–50.

Comisión Nacional de Hidrocarburos. "Análisis de Reservas de Hidrocarburos 1P, 2P y 3P. Al 1 de enero de 2021." April 1, 2021. https://www.gob.mx/cms/uploads/attachment/file/631695/2021.04.20._DSD_-_OdG_Reservas_al_1-ene-2021._vf-web-CNH.pdf.

Coronil, Fernando. *The Magical State: Nature, Money, and Modernity in Venezuela*. Chicago: University of Chicago Press, 1997.

Costilla-Salazar, Rogelio, Antonio Trejo-Acevedo, Diana Rocha-Amador, Octavio Gaspar-Ramírez, Fernando Díaz-Barriga, and Iván Nelinho Pérez-Maldonado. "Assessment of Polychlorinated Biphenyls and Mercury Levels in Soil and

Biological Samples from San Felipe, Nuevo Mercurio, Zacatecas, Mexico." *Bulletin of Environmental Contamination and Toxicology* 86, no. 2 (2010): 212–16. https://doi.org/10.1007/s00128-010-0165-z.

Crosby, Alfred W. *Ecological Imperialism: The Biological Expansion of Europe, 900–1900*. Cambridge: Cambridge University Press, 2015.

Crowder-Taraborrelli, Tomás. "Migration, Regional Traditions, and the Intricacy of Documentary Representation in *Cuates de Australia* and *La chica del sur*." *Latin American Perspectives* 41, no. 1 (2013): 172–76.

Cunha, Daniel. "The Geology of the Ruling Class?" *Anthropocene Review* 2, no. 3 (2015): 262–66. https://doi.org/10.1177/2053019615607069.

Dávila, Amparo. *Cuentos Reunidos*. Mexico DF: Fondo de Cultura Económica 2009.

———. "Espejo lento." In *Poesía reunida*. Mexico DF: Fondo de Cultura Económica, 2013.

———. *The Houseguest and Other Stories*. Translated by Audrey Harris and Matthew Gleeson. New York: New Directions, 2018.

de la Cadena, Marisol. *Earth Beings: Ecologies of Practice across Andean Worlds*. Durham, NC: Duke University Press, 2015.

———. "Uncommoning Nature." *e-flux* 65 (May 2015): 1–8.

———. "An Invitation to Live Together: Making the 'Complex We.'" *Environmental Humanities* 11, no. 2 (2019): 477–84.

del Valle, Ivonne. "On Shaky Ground: Hydraulics, State Formation, and Colonialism in Sixteenth-Century Mexico." *Hispanic Review* 77, no. 2 (2009): 197–220. https://doi.org/10.1353/hir.0.0056.

———. "From José De Acosta to the Enlightenment: Barbarians, Climate Change, and (Colonial) Technology as the End of History." *The Eighteenth Century* 54, no. 4 (2013): 435–59. https://doi.org/10.1353/ecy.2013.0042.

Dedina, Serge. *Saving the Gray Whale: People, Politics, and Conservation in Baja California*. Tucson: University of Arizona Press, 2000.

Delany, Samuel. *The Jewel-Hinged Jaw: Notes on the Language of Science Fiction*. Middletown: Wesleyan University Press, 2009.

Deleuze, Gilles, and Félix Guattari. *A Thousand Plateaus: Capitalism and Schizophrenia*. Translated by Brian Massumi. Minneapolis: University of Minnesota Press, 2003.

Delgado, Gian C. *Ecología política de la minería en América Latina*. Mexico: CIIH/UNAM, 2010.

DeLoughrey, Elizabeth M. *Allegories of the Anthropocene*. Durham, NC: Duke University Press, 2019.

Demaria, Federico, Francois Schneider, Filka Sekulova, and Joan Martinez-Alier. "What Is Degrowth?: From an Activist Slogan to a Social Movement." *Environmental Values* 22, no. 2 (2013): 191–215. https://doi.org/10.3197/0963271 13X13581561725194.

Demby, Samantha. "Mining Culture Wars Escalate in Oaxaca." NACLA, December 2, 2019. https://nacla.org/news/2019/12/01/mining-culture-wars-oaxaca-tourism.

Derrida, Jacques. *Acts of Literature*. Edited by Derek Attridge. New York: Routledge, 1992.

———. *Spectres of Marx: The State of the Debt, the Work of Mourning, and the New International*. Translated by Peggy Kamuf. New York: Routledge Classics, 2006.

DeVries, Scott M. *Creature Discomfort: Fauna-Criticism, Ethics, and the Representation of Animals in Spanish American Fiction and Poetry*. Leiden: Brill Rodopi, 2016.

Díaz, Floriberto. *Escrito: Comunalidad, energía viva del pensamiento mixe. ayuujktsënää'yën - ayuujkwënmää'ny - ayuujk mëkäjtën*. Edited by Sofía Robles Hernández and Rafael Cardoso Jiménez. Mexico DF: Universidad Nacional Autónoma de México, 2007.

Dolan, Jill. *Utopia in Performance: Finding Hope at the Theater*. Ann Arbor: University of Michigan Press, 2005.

Domínguez, Leonardo. "Mayan Train Puts More Than 2,000 Jaguars at Risk." *El Universal*, January 1, 2019. https://www.eluniversal.com.mx/english/mayan-train-puts-more-2000-jaguars-risk.

Dunlap, Alexander. "Wind Energy: Toward a 'Sustainable Violence' in Oaxaca." *NACLA* 49, no. 4 (2017): 483–88.

Echeverría, Bolívar. *Modernity and "Whiteness."* Translated by Rodrigo Ferreira. Cambridge: Polity, 2019.

Ejército Zapatista de Liberación Nacional. "Intervención de Marcos en la mesa 1 del Encuentro Intercontinental, 30 de julio de 1996." In *EZLN: Documentos y comunicados 3: 2 de octubre de 1995–24 de enero de 1997*. Mexico: Era, 1997.

Emmelhainz, Irmgard. "Images Do Not Show: The Desire to See in the Anthropocene." In *Art in the Anthropocene: Encounters among Aesthetics, Politics, Environments, and Epistemologies*, edited by Heather Davis and Etienne Turpin, 136–37. London: Open Humanities Press, 2015.

———. "Decolonial Love." *e-flux*, no. 99 (April 2019). https://www.e-flux.com/journal/99/262398/decolonial-love.

———. *Toxic Loves, Impossible Futures: Feminist Living as Resistance*. Nashville, TN: Vanderbilt University Press, 2022.

"En tiempos de crisis climática, el futuro es un territorio a defender.," #FuturosIndigenas, June 2021. https://futurosindigenas.org/manifiesto.

Escalante-Sandoval, Carlos, and Pedro Nuñez-Garcia. "Meteorological Drought Features in Northern and Northwestern Parts of Mexico under Different Climate Change Scenarios." *Journal of Arid Land* 9, no. 1 (2016): 65–75. https://doi.org/10.1007/s40333-016-0022-y.

Escobar, Arturo. *Designs for the Pluriverse: Radical Interdependence, Autonomy, and the Making of Worlds*. Durham, NC: Duke University Press, 2018.

Esteva, Gustavo. "El atropello redentor." *La Jornada*, June 29, 2020. https://www.jornada.com.mx/2020/06/29/opinion/018a2pol.

Estok, Simon C. "Theorising the EcoGothic." *Gothic Nature* 1 (2019): 34–53.

Estrada, Erick. "FICM 2015: La angustia y la resistencia." *Cinegarage*, October 31, 2015. https://www.cinegarage.com/37100-ficm-2015-06.

Evans, Brad, and Julian Reid. "Dangerously Exposed: The Life and Death of the Resilient Subject." *Resilience: International Policies, Practices, and Discourses* 1, no. 2 (2013): 83–98.

"EZLN rechaza Tren Maya porque destruirá su territorio; 'les falta información,' responde AMLO." Animal Político, January 2, 2020. https://www.animalpolitico.com/2020/01/ezln-rechaza-tren-maya-falta-informacion-responde-amlo.

Falicov, Tamara L. "Mobile Cinemas in Cuba: The Forms and Ideology of Traveling Exhibitions." *Public* 40 (2012).

———. "'Cine en Construcción' / 'Films in Progress': How Spanish and Latin American Film-Makers Negotiate the Construction of a Globalized Art-House Aesthetic." *Transnational Cinemas* 4, no. 2 (2013): 253–71.

———. *Latin American Film Industries*. London: BFI, 2019.

Farías, Iván. "Algunas precisiones sobre el terror en México." LJA.mx, May 27, 2012. https://www.lja.mx/2012/05/literatura-de-terror-en-mexico.

Farrier, David. *Anthropocene Poetics: Deep Time, Sacrifice Zones, and Extinction*. Minneapolis: University of Minnesota Press, 2019.

Fay, Jennifer. *Inhospitable World: Cinema in the Time of the Anthropocene*. Oxford: Oxford University Press, 2018.

"FEMSA Biennial Will Hold Its 14th Edition in Michoacán." FEMSA, press release, January 14, 2020. https://www.femsa.com/en/press-room/press-release/the-femsa-biennial-will-hold-its-14th-edition-in-michoacan.

"FEMSA Foundation." Nature Conservancy, accessed February 26, 2022. https://www.nature.org/en-us/about-us/who-we-are/how-we-work/working-with-companies/companies-investing-in-nature1/femsa.

Feng, Shuaizhang, Alan B. Krueger, and Michael Oppenheimer. "Linkages among Climate Change, Crop Yields and Mexico-US Cross-Border Migration." *Proceedings of the National Academy of Sciences* 107, no. 32 (2010): 14257–62.

Figueroa, Sebastián. "Apuntes sobre cine y extractivismo," *laFuga* 26 (2022). https://www.lafuga.cl/apuntes-sobre-cine-y-extractivismo/1100.

Fisher, Mark. *Ghosts of My Life: Writings on Depression, Hauntology, and Lost Futures*. Winchester: Zero, 2004.

Flanigan, Peter. "The Environmental Cost of Filmmaking," *UCLA Entertainment Law Review* 10, no. 1 (2002): 69–95.

Flores, Malva. "Revuelta. Veinticinco años de poesía en México." *Cuadernos hispanoamericanos* 809 (2017): 4–17.

Fornoff, Carolyn, and Gisela Heffes. *Pushing Past the Human in Latin American Cinema*. Albany: State University of New York Press, 2021.

Fornoff, Carolyn. "Mexican Cinema as Petrocinema." *Studies in Spanish and Latin American Cinemas* 18, no. 3 (2021): 377–87.

———. "Reflexive Extractivist Aesthetics." *FORMA* 2, no. 1 (2023): 37–69.

———. "Speculative Climate Change in Amado Nervo's 'Las nubes.'" *Paradoxa* 30 (2018): 15–34.

Forns-Broggi, Roberto. "Los retos del ecocine en nuestras Américas: Rastreos del buen vivir en tierra sublevada." *Revista de Crítica Literaria Latinoamericana* 40, no. 79 (2014): 315–34.

Freije, Vanessa. *Citizens of Scandal: Journalism, Secrecy, and the Politics of Reckoning in Mexico*. Durham, NC: Duke University Press, 2020.

Frouman-Smith, Erica. "Patterns of Female Entrapment and Escape in Three Short Stories by Amparo Dávila." *Chasqui* 18, no. 2 (1989).

Fuentes, Carlos. *Cristobal nonato*. Mexico DF: Fondo de Cultura Económica, 1988.

Fuentes, Marcela A. *Performance Constellations. Networks of Protest and Activism in Latin America*. Ann Arbor: University of Michigan Press, 2019.

Gago, Verónica. *Neoliberalism from Below: Popular Pragmatics and Baroque Economies*. Translated by Liz Mason-Deese. Durham, NC: Duke University Press, 2017.

Galeano, Juan Carlos, and Laura Parcés. "Introduction: Ecopoetry and Eco-Art from Latin America." *Global South* 16, no. 1 (2023): 9-12.

Gálvez, Alyshia. *Eating NAFTA: Trade, Food Policies, and the Destruction of Mexico*. Berkeley: University of California Press, 2018.

García, Betzabé, dir. *Venecia, Sinaloa*. Vimeo, 2011. https://vimeo.com/44362796.

———. *Los reyes del pueblo que no existe*. Vimeo, 2015. https://vimeo.com/122781305.

———. "Unsilenced." *New York Times* Op-Docs, 10:02, March 22, 2016. https://www.nytimes.com/2016/03/22/opinion/unsilenced.html.

García Hernández, Álvaro. "El pueblo fantasma de Nuevo Mercurio y la minería insostenible." *La Jornada Zacatecas*, July 9, 2014. http://ljz.mx/2014/07/09/el-pueblo-fantasma-de-nuevo-mercurio-y-la-mineria-insostenible.

García Manríquez, Hugo. *Anti-Humboldt: A Reading of the North American Free Trade Agreement*. Brooklyn: Litmus Press, 2015.

García-Mora, Fernando, and Jorge Mora-Rivera. "Exploring the Impacts of Internet Access on Poverty: A Regional Analysis of Rural Mexico." *New Media and Society* 25, no. 1 (Jan. 2023): 26–49. https://doi.org/10.1177/14614448211000650.

Garza, Valentina, and Juan Manuel Pérez. "La provincia minera de Zacatecas y su evolución demográfica (1700–1810)." *Historias* 77 (2010): 53–86.

Gerber Bicecci, Verónica, Canek Zapata, and Carlos Bergen. "La máquina distópica" (website). Accessed March 5, 2022. https://lamaquinadistopica.xyz.

Gerber Bicecci, Verónica. *Empty Set*. Translated by Christina MacSweeney. Minneapolis: Coffee House Press, 2018.

———. *Otro día . . . (poemas sintéticos)*. Mexico DF: Almadía, 2019.

———. *In the Eye of Bambi*. London: Whitechapel Gallery, 2020.

———. *La compania*. Mexico DF: Almadia, 2020.

———. *La tierra es plana como una hoja (y cabalga en el aire)*. Mexico City: Gato Negro, 2021. https://www.veronicagerberbicecci.net/la-tierra-es-plana-earth-is-flat.

———. "El Vocabulario b." Verónica Gerber Bicecci (website), 2019; accessed February 25, 2022. https://www.veronicagerberbicecci.net/vocabulario-b-b-vocabulary.

Ghosh, Amitav. *The Great Derangement: Climate Change and the Unthinkable.* Chicago: University of Chicago Press, 2016.

Giddens, Anthony. *Modernity and Self Identity: Self and Society in the Late Modern Age.* Cambridge, UK: Polity, 1991.

Global Witness. "Decade of Defiance," September 2022. https://www.globalwitness.org/en/campaigns/environmental-activists/decade-defiance/#a-global-analysis-2021.

———. "Standing Firm," September 2023. https://www.globalwitness.org/en/campaigns/environmental-activists/standing-firm.

Goldsmith, Ben, and Tom O'Regan. *The Film Studio: Film Production in the Global Economy.* Lanham, MD: Rowman and Littlefield, 2005.

Gollnick, Brian. *Reinventing the Lacandón: Subaltern Representations in the Rain Forest of Chiapas.* Tucson: University of Arizona Press, 2008.

Gomez, Rocio. *Silver Veins, Dusty Lungs: Mining, Water, and Public Health in Zacatecas, 1835–1946.* Lincoln: University of Nebraska Press, 2020.

Gómez-Barris, Macarena. *The Extractive Zone: Social Ecologies and Decolonial Perspectives.* Durham, NC: Duke University Press, 2017.

González de Arce Arzave, Rocío Betzabeé. "El viaje del cine mexicano de ficción hacia la conciencia ecológica: Imaginarios de la naturaleza, ecoutopías y ética ambiental en la pantalla." MA thesis, Universidad Iberoamericana, 2019.

Gonzalez, Anita. "Indigenous Acts: Black and Native Performances in Mexico." *Radical History Review* 2009, no. 103 (2009): 131–41.

González, Everardo, dir. *Cuates de Australia.* Mexico DF: Ciénaga, 2013.

González-Flores, Laura. "What Is Present, What Is Visible: The Photo-Portraits of the 43 'Disappeared' Students of Ayotzinapa as Positive Social Agency." *Journal of Latin American Cultural Studies* 27, no. 4 (2018): 487–506.

Gottesdiener, Laura. "Dams, Taps Running Dry in Northern Mexico amid Historic Water Shortages," Reuters, June 20, 2022. https://www.reuters.com/world/americas/dams-taps-running-dry-northern-mexico-amid-historic-water-shortages-2022-06-20.

Grosfoguel, Ramón. "Del 'extractivismo económico' al 'extractivismo ontológico': Una forma destructiva de conocer, ser y estar en el mundo." *Tabula Rasa* 24 (2016): 123–43.

Guerrero, Maricela. *El sueño de toda célula.* Mexico DF: Ediciones Antílope, 2018.

Gutiérrez, Laura G. "*Resiliencia Tlacuache* by Naomi Rincón Gallardo." *Terremoto*, May 11, 2020. https://terremoto.mx/en/online/resilencia-tlacuache/.

Haraway, Donna. *Staying with the Trouble: Making Kin in the Chthulucene.* Durham, NC: Duke University Press, 2016.

Hartman, Sadiya. "Venus in Two Acts." *Small Axe: A Caribbean Journal of Criticism* 12, no. 2 (2008): 1–14.

Heffes, Gisela. *Políticas de la destrucción / Poéticas de la preservación: Apuntes para una lectura (eco)crítica del medio ambiente en América Latina*. Rosario: Beatriz Viterbo, 2013.

———. "Estéticas del antropoceno: Tres preguntas con Isabel Zapata." Hablemos escritoras, December 27, 2021. https://www.hablemosescritoras.com/posts/719.

———. "Toxicity." In *Handbook of Latin American Environmental Aesthetics*, edited by Jens Andermann, Gabriel Giorgi, and Victoria Saramago, 395–420. Berlin: De Gruyter, 2023.

Heise, Ursula K. *Imagining Extinction: The Cultural Meanings of Endangered Species*. Chicago: University of Chicago Press, 2016.

Henríquez, Elio. "Piden revocar permiso de agua a FEMSA en Chiapas." *La Jornada*, June 25, 2020. https://www.jornada.com.mx/2020/06/25/politica/017n1pol.

Hernández Ibarzábal, José Alberto, and David Bonilla. "Examining Mexico's Energy Policy under the 4T." *Extractive Industries and Society* 7, no. 2 (2020): 669–75. https://doi.org/10.1016/j.exis.2020.03.002.

Hernández, Bernadine. "Living on All Fours: Latinx Performance and the Trans Human Turn in *Cuatro Patas*." *Transgender Studies Quarterly* 6, no. 2 (January 2019): 260–68.

Herrero Garvín, Laura, dir. *El Remolino*. Instituto Mexicano de Cinematografía, 2016.

Higashi, Alejandro. "Hitos provisionales en el perfil de una generación: Poetas mexicanos nacidos entre 1975 y 1985." *Literatura Mexicana* 25, no. 2 (2014): 49–74.

———. "Karen Villeda habla sobre *Dodo*." *Ancila: Crítica de Poesía Mexicana Contemporánea*, no. 2 (June 2014)

Hind, Emily. *Dude Lit: Mexican Men Writing and Performing Competence, 1955–2012*. Tucson: University of Arizona Press, 2019.

———. "*Provincia* in Recent Mexican Cinema, 1989–2004," *Discourse* 26, no. 1–2 (Winter and Spring 2004): 26–45.

Hoge, Charles. "The Dodo in the Long Eighteenth Century: An Exploration of the Gray Ghost outside of the English Sentimental Eye." *University of Toronto Quarterly* 83, no. 3 (2014): 687–704.

Hoyos, Héctor. *Things with a History: Transcultural Materialism and the Literatures of Extraction in Contemporary Latin America*. New York: Columbia University Press, 2019.

Iheka, Cajetan. *African Ecomedia: Network Forms, Planetary Politics*. Durham, NC: Duke University Press, 2021.

INEGI. "Encuesta Nacional de Ingresos y Gastos de los Hogares (ENIGH): 2012 Nueva construcción." Accessed March 3, 2022. https://www.inegi.org.mx/programas/enigh/nc/2012/#Documentacion.

———. "Comunicado de prensa," news release no. 352/21, June 22, 2021.

———. "Población rural y urbana." Cuéntame de Mexico, 2020, accessed February 2, 2022. http://cuentame.inegi.org.mx/poblacion/rur_urb.aspx?tema=P.

IUCN Red List. "Summary Statistics 2019." https://www.iucnredlist.org/resources/summary-statistics.

Jakobson, Roman. *Studies on Child Language and Aphasia*. The Hague: Mouton, 1971.

Janzen, Rebecca. "*El Cambio / The Change* Joskowicz, ([1971] 1975): Mexican Counterculture and the Futility of Protest in the 1970s." *Studies in Spanish and Latin American Cinema* 18, no. 2 (January 2021): 159–75.

JM, Rodolfo. "13 ideas acerca de la literatura de terror Mexicana." LJA.mx, May 27, 2012. https://www.lja.mx/2012/05/literatura-de-terror-en-mexico.

"Juez concede libertad a los responsables identificados de la desaparición forzosa de Sergio Rivera." Frontline Defenders, September 15, 2020. https://www.frontlinedefenders.org/es/case/disappearance-sergio-rivera-hernandez#case-update-id-12392.

Kartha, Sivan, Eric Kemp-Benedict, Emily Ghosh, and Anisha Nazareth. *The Carbon Inequality Era: An Assessment of the Global Distribution of Consumption Emissions among Individuals from 1990 to 2015 and Beyond*. Stockholm: Stockholm Environment Institute and Oxfam International, 2020.

Keizman, Betina. "Territorios y naturaleza bajo la transmutación del archivo." *Valenciana* 12, no. 24 (2019): 229–45. https://doi.org/10.15174/rv.v0i24.473.

Kendrick, Anna Kathryn. "Homero Aridjis, Public Pedagogy and an Educational Poetics of Environmentalism." *Bulletin of Hispanic Studies* 100, no. 1 (2023): 97–116.

Kierkegaard, Soren. *Papers and Journals*. Translated by Alastair Hannay. New York: Penguin Books, 2015.

Klein, Cecilia F. "Fighting with Femininity: Gender and War in Aztec Mexico." *Estudios de cultura náhuatl* 24 (1994): 219–53.

Knox-Russell, Allyse. "Futurity without Optimism: Detaching from Anthropocentrism and Grieving Our Fathers in *Beasts of the Southern Wild*." In *Affective Ecocriticism: Emotion, Embodiment, Environment*, edited by Kyle Bladow and Jennifer Ladino, 213–32. Lincoln: University of Nebraska Press, 2018.

Koch, Alexander, Chris Brierley, Mark M. Maslin, and Simon L. Lewis. "Earth System Impacts of the European Arrival and Great Dying in the Americas after 1492." *Quaternary Science Reviews* 201 (2019): 13–36.

Kristeva, Julia. "Word, Dialogue and Novel." In *The Kristeva Reader*, edited by Toril Moi, 34–61. New York: Columbia University Press, 1986.

Kulp, Scott A., and Benjamin H. Strauss. "New Elevation Data Triple Estimates of Global Vulnerability to Sea-Level Rise and Coastal Flooding." *Nature Communications* 10 (2019). https://doi.org/10.1038/s41467-019-12808-z.

Ladino, Jennifer. *Reclaiming Nostalgia: Longing for Nature in American Literature*. Charlottesville: University of Virginia Press, 2012.

LaDuke, Winona, and Deborah Cowen. "Beyond Wiindigo Infrastructure." *South Atlantic Quarterly* 119, no. 2 (2020): 243–68. https://doi.org/10.1215/00382876-8177747.

Latour, Bruno. *We Have Never Been Modern*. Translated by Catherine Porter. Cambridge: Harvard University Press, 1993.

Law, John. "What's Wrong with a One-World World?" *Distinktion: Journal of Social Theory* 16, no. 1 (2015): 126–39.

Leary, John Patrick. "Keywords for the Age of Austerity 19: Resilience." Keywords: The New Language of Capitalism, June 23, 2015, https://keywordsforcapitalism.com/2015/06/23/keywords-for-the-age-of-austerity-19-resilience.

Ledesma, Eduardo. *Radical Poetry: Aesthetics, Politics, Technology, and the Ibero-American Avant-Gardes, 1900–2015*. Albany, NY: SUNY Press, 2016.

Lefebvre, Henri. *The Production of Space*. Translated by Donald Nicholson-Smith. Malden, MA: Blackwell, 1991.

———. "Space and the State." In *State, Space, World*. Translated by Gerald Moore, Neil Brenner, and Stuart Elden, 223–53. Minneapolis: University of Minnesota Press, 2009.

Leff, Enrique. *Ecología y capital: Racionalidad ambiental, democracia participativa y desarrollo sustentable*, 6th ed. México: Siglo XXI, 2005.

Lehner, Flavio, Sloan Coats, Thomas F. Stocker, Angeline G. Pendergrass, Benjamin M. Sanderson, Christoph C. Raible, and Jason E. Smerdon. "Projected Drought Risk in 1.5°C and 2°C Warmer Climates." *Geophysical Research Letters* 44, no. 14 (2017): 7419–28.

Levinas, Emmanuel. *Totality and Infinity: An Essay on Exteriority*. Translated by Alphonso Lingis. Pittsburgh, PA: Duquesne University Press, 1969.

———. "Language and Proximity." In *Collected Philosophical Papers*, 109–26. Translated by Alphonso Lingus. Pittsburgh, PA: Duquesne University Press, 1998.

Lewis, Simon L., and Mark A. Maslin. "Defining the Anthropocene." *Nature* 519 (2015).

Lifshey, Adam. *Specters of Conquest: Indigenous Absence in Transatlantic Literatures*. New York: Fordham University Press, 2010.

Lindero, Scarlett. "Pemex contaminó 655 lugares en México entre 2008 y 2021: Semarnat." *Gatopardo*, August 15, 2022. https://gatopardo.com/noticias-actuales/pemex-medio-ambiente-danos.

Llamas-Rodriguez, Juan. "A Global Cinematic Experience: Cinépolis, Film Exhibition, and Luxury Branding." *JCMS: Journal of Cinema and Media Studies* 58, no. 3 (2019): 49–71.

Llano Vázquez Prado, Manuel, and Carla Flores Lot. *La contribución de Pemex a la emergencia climática: Análisis de emisiones por campo petrolero desde 1960*. Mexico DF: CartoCrítica, 2019.

Lomnitz, Claudio. "Afterword: Spread It Around!" In *Cultural Agency in the Americas*, edited by Doris Sommer, 334–40. Durham, NC: Duke University Press, 2006.

———. *La nación desdibujada: México en trece ensayos*. Mexico DF: Malpaso, 2016.

López Austin, Alfredo. *Los mitos del tlacuache: Caminos de la mitología mesoamericana*. Mexico DF: Universidad Autónoma de México, 1996.

Loría Araujo, David, and Francisco G. Tijerina Martínez. "La crisis del capital en dos energoficciones contemporáneas: *Temporada de huracanes*, de Fernanda Melchor y *La compañía*, de Verónica Gerber Bicecci." *De Raíz Diversa* 9, no. 17 (2022): 121–47.

Lozano Herrera, Rubén. *Las veras y las burlas de José Juan Tablada*. Mexico DF: Universidad Iberoamericana, 1995.

Luna Chávez, Marisol, and Víctor Díaz Arciniega. "La rutina doméstica como figuración siniestra. Amparo Dávila: Su poética del dolor." *Sincronía* 74 (2018): 205–33.

Lustgarten, Abrahm. "Where Will Everyone Go?" ProPublica, July 23, 2020. https://features.propublica.org/climate-migration/model-how-climate-refugees-move-across-continents.

MacDonald, Scott. *American Ethnographic Film and Personal Documentary: The Cambridge Turn*. Berkeley: University of California Press, 2013.

Machado, Horacio. *Minar: Colonialidad y genealogía del extractivismo*, 3rd ed. Mexico DF: Ediciones OnA, 2020.

Maciel Cabañas, Pamela. "El lobo mexicano." In "Extinción," *Revista de la Universidad de México*, November 2017. https://www.revistadelauniversidad.mx/articles/e3ab271b-74cf-43d7-b351-bf8cc8450ee4/el-lobo-mexicano.

Malamud, Randy. "'José Emilio Pacheco: 'I Saw a Dying Fish.'" In *Poetic Animals and Animal Souls*, 77–92. New York: Palgrave, 2003.

Mantecón, Ana Rosas. *Ir al cine: Antropología de los públicos, la ciudad y las pantallas*. Mexico DF: Universidad Autónoma Metropolitana, 2017.

Marán, Luna. "Piel territorio," Tzam Trece Semillas, July 2021, https://tzamtrecesemillas.org/sitio/piel-territorio.

Marcone, Jorge. "Filming the Emergence of Popular Environmentalism in Latin America: Postcolonialism and Buen Vivir." In *Global Ecologies and the Environmental Humanities: Postcolonial Approaches*, edited by Elizabeth DeLoughrey, Jill Didur, and Anthony Carrigan, 207–24. New York: Routledge, 2015.

Marks, Laura U., and Radek Przedpelski. "Bandwidth Imperialism and Small-File Media." Post45, April 13, 2021. https://post45.org/2021/04/bandwidth-imperialism-and-small-file-media.

Martínez Martínez, Germán."El cine de Everardo González y las posibilidades del documental en México." In *Reflexiones sobre cine mexicano contemporáneo: Documental*, edited by Claudia Curiel de Icaza and Abel Muñoz Hénonin. Mexico DF: Cineteca Nacional, 2014.

Martínez Ranero, Víctor. Review of *Los reyes del pueblo que no existe*, Time Out, October 20, 2015. https://www.timeout.com/es/cine/los-reyes-del-pueblo-que-no-existe.

Marx, Karl. *Capital: A Critique of Political Economy*, vol. 1. Translated by Samuel Moore and Edward Aveling. Mineola: Dover, 2011.

Mason, Margaret L., and Yulan M. Washburn. "The Bestiary in Contemporary Spanish American Literature." *Revista de Estudios Hispánicos* 8, no. 2 (1974).

Matías Rendón, Ana. "Contra el despojo." *Ojarasca* 264 (April 2019): 3–6. http://ojarasca.jornada.com.mx/2019/04/12/contra-el-despojo-264-8847.html.

Mauri, Caterina, and Andrea Sansó. "The Linguistic Marking of (Ir)Realis and Subjunctive." In *The Oxford Handbook of Modality and Mood*, edited by Jan Nuyts and Johan Van Der Auwera. Oxford, UK: Oxford University Press, 2016.

Maza, Santiago, dir. *El Tema*. Episode 4, "Energía." Aired May 3, 2021. https://lacorrientedelgolfo.net/proyecto/el-tema/energia.

Mazza de Luca, Emiliano, dir. *Nueva Venecia*. Montevideo: Passaparola, 2016.

Mbembe, Achille. *Necropolitics*. Translated by Steve Corcoran. Durham, NC: Duke University Press, 2019.

McClennen, Sophia. *Globalization and Latin American Cinema: Toward a New Critical Paradigm*. London: Palgrave Macmillan, 2018.

McKay, Micah. "'Pasto sin fin del basurero': Trash and Disposal in the Poetry of José Emilio Pacheco." *Latin American Literary Review* 47, no. 93 (2020): 49–58.

McKay, Robert R. "Representation." In *Critical Terms for Animal Studies*, edited by Lori Gruen, 307–319. Chicago: University of Chicago Press, 2018.

Méndez Cota, Gabriela. "Policing the Environmental Conjecture: Structural Violence in Mexico and the National Assembly of the Environmentally Affected." *New Formations* 96/97 (2019): 69–88.

Mendoza, Natalia. "2014: Mancha naranja de trescientos kilómetros." In *1968–2018: Historia colectiva de medio siglo*, edited by Claudio Lomnitz. Mexico DF: Universidad Nacional Autónoma de México, 2018.

Mendoza-Hernández, J. María, Alejandro Zermeño-González, J. Manuel Covarrubias-Ramírez, and J. Jesús Cortés-Bracho. "Proyecciones climáticas para el estado de Coahuila usando el modelo precis bajo dos escenarios de emisiones." *Agrociencia* 47 (2013): 523–37.

Miller, Todd. *Empire of Borders: The Expansion of the US Border around the World*. New York: Verso, 2019.

Ministry of Environment and Natural Resources. "Nationally Determined Contributions: 2020 Update." Government of Mexico, 2020. https://www4.unfccc.int/sites/ndcstaging/PublishedDocuments/Mexico%20First/NDC-Eng-Dec30.pdf.

Mirzoeff, Nicholas. "It's Not the Anthropocene, It's the White Supremacy Scene; or, The Geological Color Line." In *After Extinction*, edited by Richard Grusin, 123–49. Minneapolis: University of Minnesota Press, 2018.

Moore, Jason W. "Ecology, Capital, and the Nature of Our Times: Accumulation and Crisis in the Capitalist World-Ecology." *Journal of World Systems Research* 17, no. 1 (2011): 107–46.

Monsiváis, Carlos. *Mexican Postcards*. Translated by John Kraniauskus. London: Verso, 1997.

———. *"No sin nosotros": Los días del terremoto 1985–2005*. Mexico DF: Era, 2005.

Mora, Mariana. "Desparición forzada, racismo institucional y pueblos indígenas en el caso Ayotzinapa, México." *LASA Forum* 48, no. 2 (2017): 29–30.

Moraru, Christian. *Rewriting: Postmodern Narrative and Cultural Critique in the Age of Cloning*. Albany, NY: SUNY Press, 2001.

Moraga, Cherríe. *The Last Generation: Prose and Poetry*. Boston: South End Press, 1993.

Morla, Jorge. "Marichuy: 'De los indígenas, al Gobierno mexicano solo le interesa el folclore.'" *El País*, October 12, 2019. https://elpais.com/internacional/2019/10/12/actualidad/1570904742_707998.html.

"Muere trabajador en accidente en la mina de San José del Progreso." *La Minuta* (blog), EDUCA, August 20, 2019. https://www.educaoaxaca.org/muere-trabajador-en-accidente-en-la-mina-de-san-jose-del-progreso.

Murphy, Michelle. "Alterlife and Decolonial Chemical Relations." *Cultural Anthropology* 32, no. 4 (2017): 494–503.

Murray-Tortarolo, Guillermo N. "Seven Decades of Climate Change across Mexico." *Atmósfera* 34, no. 2 (2021): 217–26.

Ndalianis, Angela. *The Horror Sensorium: Media and the Senses*. Jefferson, NC: McFarland, 2012.

"NAFTA's Impact on North American Hazardous Waste Imports and Exports." *Hazardous Waste Consultant* 19, no. 3 (2001): 2.16–2.19.

Navarrete, Federico. "El lugar de las siete cuevas." *Revista de la Universidad de México* 80 (February 2019).

Navarro Trujillo, Mina Lorena. *Luchas por lo común: Antagonismo social contra el despojo capitalista de los bienes naturales en México*. Mexico DF: Bajo Tierra, 2015.

Negrín, Edith. *Letras sobre un dios mineral: El petróleo mexicano en la narrativa*. Mexico DF: El Colegio de México, 2017.

Neira, Hidalgo. "Cine Móvil ToTo llevará cine, de manera sustentable, más allá de las grades ciudades." *Reporte Indigo*, August 18, 2021. https://www.reporteindigo.com/piensa/cine-movil-toto-llevara-cine-de-manera-sustentable-mas-alla-de-las-grandes-ciudades.

Nepote, Mónica. "Humanos y no humanos: Lenguajes emitidos, pensamientos escuchados." *Las repúblicas de lo salvaje* (blog), August 18, 2022. https://lasrepublicasdelosalvaje.blog/blog/humanos-y-no-humanos-lenguajes-emitidos-pensamientos-escuchados.

Nesselrode Moncada, Sean. *Refined Material: Petroculture and Modernity in Venezuela*. Oakland, CA: University of California Press, 2023.

Nichols, Bill. *Representing Reality: Issues and Concepts in Documentary*. Bloomington: Indiana University Press, 1991.

Nixon, Rob. *Slow Violence and the Environmentalism of the Poor*. Cambridge, MA: Harvard University Press, 2013.

Noya, Dení. "Minera Cuzcatlán y Oaxaca FilmFest, crónica de un boicot anunciado." Avispa, October 12, 2019. https://avispa.org/minera-cuzcatlan-y-oaxaca-filmfest-cronica-de-un-boicot-anunciado.

Oloff, Kerstin. "The 'Monstrous Head' and the 'Mouth of Hell': The Gothic Ecologies of the 'Mexican Miracle.'" In *Ecological Crisis and Cultural Representation in Latin America: Ecocritical Perspectives on Art, Film, and Literature*, edited by Mark Anderson and Zélia M. Bora, 79–98. Lanham, MA: Lexington, 2016.

Ordóñez, Samantha. *Mexico Unmanned: The Cultural Politics of Masculinity in Mexican Cinema*. Albany, NY: SUNY Press, 2022.

Ortega, Pedro. "Betzabé García: *Los reyes del pueblo que no existe*." FilmArte, 2015. https://web.archive.org/web/20170508163611/http://www.filmarte.net/Entrevistas/betzabe-garcia-los-reyes-del-pueblo-que-no-existe.

Osorio, Jaime. "Crisis estatales y violencia desnuda: La excepcionalidad mexicana." In *Violencia y crisis del estado: Estudio sobre México*, edited by Jaime Osorio, 33–62. Mexico DF: Universidad Autónoma Metropolitana, 2011.

———. *Reproducción del capital, estado y sistema mundial: Estudios desde la teoría marxista de la dependencia*. Bogotá, DC: Universidad Nacional de Colombia, 2017. https://www.uneditorial.com/bw-reproduccion-del-capital-estado-y-sistema-mundial-estudios-desde-la-teoria-marxista-de-la-dependencia.html.

Ota, Seiko. "José Juan Tablada: La influencia del haikú japonés en *Un día* . . . " *Literatura Mexicana* 16, no. 1 (2005): 133–44.

Oyedotun, Temitope D. Timothy, Arturo Ruiz-Luna, and Alma G. Navarro-Hernández. "Contemporary Shoreline Changes and Consequences at a Tropical Coastal Domain." *Geology, Ecology, and Landscapes* 2, no. 2 (2018): 104–14.

Pacheco, José Emilio. *Islas a la deriva: Poemas, 1973–1975*. México DF: Era, 2006.

———. *Nuevo álbum de zoología*. Mexico DF: Era, 2013.

Pacheco-Vega, Raúl. "Agua embotellada en México: De la privatización del suministro a la mercantilización de los recursos hídricos." *Espiral* 22, no. 63 (2015): 221–63.

Padilla, Ignacio. *Arte y olvido del terremoto*. México: Almadía, 2010.

Paley, Dawn. *Drug War Capitalism*. Edinburgh: AK Press, 2015.

Parikka, Jussi. "New Materialism as Media Theory: Medianatures and Dirty Matter." *Communication and Critical/Cultural Studies* 9, no. 1 (2012): 95–100.

Partido Verde Ecologista de México. "Estatutos," 2010. https://www.partidoverde.org.mx/transparencia/II/Estatutos.pdf.

Pascua Canelo, Marta. "La mirada borrosa: Poéticas del desenfoque y visiones oblicuas en la narrativa hispánica contemporánea." *Catedral Tomada* 7, no. 13 (2019): 178–201.

Paz, Rafael. "'Los reyes del pueblo que no existe' y las visiones del apocalipsis." *Butaca Ancha*, November 30, 2015. http://butacaancha.com/los-reyes-del-pueblo-que-no-existe-y-las-visiones-del-apocalipsis.

Pennington, Eric. "Amparo Dávila's 'El huésped' and Domestic Violence." *Grafemas*, December 2007, http://people.wku.edu/inma.pertusa/encuentros/grafemas/diciembre_07/pennington.html.

Pérez, Oscar A. "Local Landscapes, Global Conversations: The Case of Three Environmental Documentary Films from the Hispanic World." *Hispanic Issues On Line*, no. 24 (2019): 66–79.

Pérez Alfonso, Jorge A., and Mónica Mateos-Vega. "Construir el Tren Maya 'Va a ser un desastre,' alerta Francisco Toledo." *La Jornada*, February 6, 2019. https://www.jornada.com.mx/2019/02/06/cultura/a03n1cul.

Pérez Alfonso, Jorge A. "Borran mural de arte urbano en el centro histórico de Oaxaca." *La Jornada*, January 29, 2016. https://www.jornada.com.mx/2016/01/29/cultura/a03n1cul.

Pérez Limón, Lilia Adriana. "Documenting Precarity and Other Ghostly Remains: Passivity as Political Practice in Betzabé García's Documentary *Los reyes del pueblo que no existe*." *Studies in Latin American Popular Culture* 36 (2018): 33–46.

Plumwood, Val. "Shadow Places and the Politics of Dwelling." *Australian Humanities Review* 44 (2008): 139–50.

Prádanos, Luis I. *Postgrowth Imaginaries: New Ecologies and Counterhegemonic Culture in Post-2008 Spain*. Liverpool, UK: Liverpool University Press, 2018.

Pratt, Mary Louise. *Imperial Eyes: Travel Writing and Transculturation*. London: Routledge, 1992.

Price, Brian L. "Heterotemporal *Mise-en-scene* in the Films of Luis Estrada." *Arizona Journal of Hispanic Cultural Studies* 16 (2012): 259–74.

Price, Rachel. *Planet/Cuba: Art, Culture, and the Future of the Island*. New York: Verso, 2015.

Pskowski, Martha. "Mexico's Fracking Impasse." *NACLA*, October 27, 2020. https://nacla.org/news/2020/10/22/mexico-fracking-impasse.

Puar, Jasbir K., and Andrew Ross. "Decolonising the Museum." *Al Jazeera*, July 21, 2021. https://www.aljazeera.com/opinions/2021/7/21/decolonising-the.

"Quienes somos." Aquí cine (website), accessed February 24, 2022. http://aquicineoaxaca.blogspot.com/p/quienes-somos.html.

Quintana-Navarrete, Jorge. "José Vasconcelos's Plant Theory: The Life of Plants, Botanical Ethics, and the Cosmic Race." *Hispanic Review* 89, no. 1 (2021): 69–92.

———. "Reading Race in Rocks: Political Geology in Nineteenth-Century Mexico." *Journal of Latin American Cultural Studies* 30, no. 4 (2021): 525–43.

Quintana Solórzano, Fausto, ed. *Sociedad global, crisis ambiental y sistemas socio-ecológicos*. Mexico DF: UNAM, 2019.

Ramírez, Erick. "Disruptores: Cine Móvil ToTo, la pantalla nómada." *El Sol de México*, March 12, 2020.

Ramírez-Miranda, Francisco Javier. "Aproximaciones a lo real en el cine de Everardo González." *Comunicación y Medios* 36 (2017): 33–42.

Rangan, Pooja. *Immediations: The Humanitarian Impulse in Documentary.* Durham, NC: Duke University Press, 2017.

"Red mexicana de afectadas/os por la minería, 'Bety Cariño, nos arrebataron tu presencia y hasta hoy la impunidad impera.'" Facebook, April 27, 2021. https://www.facebook.com/REMAMX/photos/pcb.5476222835753242/5476222 549086604/.

Rediker, Marcus. *Outlaws of the Atlantic: Sailors, Pirates, and Motley Crews in the Age of Sail.* Boston, MA: Beacon, 2014.

Reyes, Alfonso. *Visión de Anáhuac (1519).* In *Obras completas de Alfonso Reyes II.* Mexico DF: Fondo de Cultura Económica, 1995.

———. "Palinodia del polvo." *Caelum* 6 (2014).

Reyes Escutia, Felipe, ed. *Construir un NosOtros con la tierra: Voces latinoamericanos por la descolonización del pensamiento y la acción ambientales.* Chiapas: Universidad de Ciencias y Artes de Chiapas, 2018.

Rigaud, Kanta Kumari, Alex de Sherbinin, Bryan Jones, Jonas Bergmann, Viviane Clement, Kayly Ober, Jacob Schewe, Susana Adamo, Brent McCusker, Silke Heuser, and Amelia Midgley. *Groundswell: Preparing for Internal Climate Migration.* Washington, DC: World Bank, 2018.

Rincón Gallardo, Naomi. "*The Formaldehyde Trip:* A Mythical/Critical (Under)world-Making Dedicated to Bety Carino," *Critical Ethnic Studies* 4, no. 2 (2018): 39–74.

———. *El viaje de formol.* Naomi Rincón Gallardo (website), 2017. https://www.naomirincongallardo.net/viaje-de-formol.html.

———. *Sangre pesada.* Naomi Rincón Gallardo (website), Bienal FEMSA, 2018. https://www.naomirincongallardo.net/sangre-pesada.html.

———. *Resiliencia Tlacuache.* Parallel Oaxaca, July 13–September 21, 2019.

Río, Fernarda. *Manual para exhibicionistas: Cómo montar un cine.* Mexico DF: IMCINE, 2015.

Riofrancos, Thea. "What Green Costs." *Logic Magazine,* December 7, 2019. https://logicmag.io/nature/what-green-costs.

———. *Resource Radicals: From Petro-Nationalism to Post-Extractivism in Ecuador.* Durham, NC: Duke University Press, 2020.

Rivera Garza, Cristina. *Grieving: Dispatches from a Wounded Country.* Translated by Sarah Booker. New York: Feminist Press, 2020.

———. *The Restless Dead: Necrowriting and Disappropriation.* Translated by Robin Myers. Nashville, TN: Vanderbilt University Press, 2020.

———. *El invencible verano de Liliana.* México: Random House, 2021.

Rodríguez, Ana Mónica. "*Bosque de niebla* muestra 'otra forma de relaciones humanas, de vivir en unidad.'" *La Jornada,* February 15, 2018. https://www.jornada.com.mx/2018/02/15/espectaculos/a09n1esp.

Rodríguez Fuentes, Óscar Daniel. "La lucha por la opinion pública en México: Sociedad civil vs. Partido Verde Ecologista de México." *Inciso* 18, no. 2 (2016): 21–35.

Rodríguez Mendoza, Xitlalitl. *Jaws (Tiburón)*. Guadalajara: Mantis Editores, 2015.

Rogers, Charlotte. *Mourning El Dorado: Literature and Extractivism in the Contemporary American Tropics*. Charlottesville: University of Virginia Press, 2019.

Rojas, Marisol. "The Multidimensional and the Multiple in Contemporary Art: *Let's Sow Dreams, Let's Harvest Hope* (2015–2018) by Lapiztola Stencil." Marisol Rojas / portfolio, accessed August 2, 2021. http://marisolrojas.com/textos_2.html.

Romero Rivera, Marcela. "Signs of the Inhuman: Hauntings and Lost Futures in Verónica Gerber Bicecci's *La Compañía*." *CLCWeb: Comparative Literature and Culture* 24, no. 1 (2022). https://doi.org/10.7771/1481-4374.4296.

Romney, Jonathan. "In Search of Lost Time." *Sight and Sound* 20, no. 2 (2010): 43–44.

Rosenthal, Lecia. *Mourning Modernism: Literature, Catastrophe, and the Politics of Consolation*. New York: Fordham University Press, 2011.

Rothberg, Michael. *The Implicated Subject: Beyond Victims and Perpetrators*. Redwood City, CA: Stanford University Press, 2019.

Ruiz Monroy, Julio E. "*Dodo* la supervivencia de la poesía extinta." *Luvina* 76 (August 2014). https://luvina.com.mx/dodoo-la-supervivencia-de-la-poesia-extinta-julio-e-ruiz-monroy.

Rust, Stephen, and Salma Monani. "Introduction: Cuts to Dissolves—Defining and Situating Ecocinema Studies." In *Ecocinema Theory and Practice*, edited by Stephen Rust, Salma Monani, and Sean Cubitt, 1–13. New York: Routledge, 2013.

Saldaña-Durán, Claudia E., Gerardo Bernache Pérez, Sara Ojeda-Benitez, and Samantha E. Cruz-Sotelo. "Environmental Pollution of E-waste: Generation, Collection, Legislation, and Recycling Practices in Mexico." In *Handbook of Electronic Waste Management: International Best Practices and Case Studies*, edited by Majeti Narasimha Vara Prasad, Meththika Vithanage, and Anwesha Borthakur. Oxford: Butterworth-Heinemann, 2020.

Sánchez, Griselda. *Aire no te vendas: La lucha por el territorio desde las ondas*. Mexico DF: IWGIA, 2017.

Sánchez, Mikeas. "Jujtzyere' / Cuánto vale?" In *Cómo ser un buen salvaje. Jujtzye tä wäpä tzamapänh'ajä*, 83–84. Guadalajara: Universidad de Guadalajara, 2019.

———. "Me sumo a los que llaman 'vándalos' de las normales rurales." *Ojarasca* 290 (June 2021): 6–7. https://www.jornada.com.mx/2021/06/12/ojarasca290.pdf.

Sánchez Matías, Nuria Angelica. "Del desastre a los futuros habitables: La visualidad, la medialidad y la textualidad en la obra ecosocial de Verónica Gerber Bicecci," MA thesis, Universidad Iberoamericana, 2023.

Sánchez Prado, Ignacio. *Screening Neoliberalism: Transforming Mexican Cinema, 1988–2012*. Nashville, TN: Vanderbilt University Press, 2014.

———. "The Politics-Commodity: The Rise of Mexican Commercial Documentary in the Neoliberal Era." In *Latin American Documentary Film in the New Millennium*, edited by María Guadalupe Arenillas and Michael J. Lazzarra, 97–114. New York: Palgrave, 2016.

———. *Strategic Occidentalism: On Mexican Fiction, the Neoliberal Book Market, and the Question of World Literature*. Evanston, IL: Northwestern University Press, 2018.

———. "Lengua precaria: La poesía mexicana en crisis epistémica." *América sin nombre* 23 (2018): 49–58.

Sandoval, Alicia. "*Otro día . . . (poemas sintéticos)* y *La compañía* de Verónica Gerber Bicecci*." *Revista de la Universidad de México*, no. 855/856 (January 2020): 151.

Sandvig, Christian. "The Internet as the Anti-Television: Distribution Infrastructure as Culture and Power." In *Signal Traffic: Critical Studies of Media Infrastructures*, edited by Lisa Parks and Nicole Starosielski, 225–45. Champaign: University of Illinois Press, 2015.

Santillán Vera, Mónica, and Angel de la Vega Navarro. "Do the Rich Pollute More?: Mexican Household Consumption by Income Level and CO_2 Emissions." *International Journal of Energy Sector Management* 13, no. 3 (2019): 694–712.

Santos, Milton. "The Return of the Territory." In *Milton Santos: A Pioneer in Critical Geogrpahy from the Global South*, edited by Lucas Melcaço and Carolyn Prouse. Cham, Switzerland: Springer, 2017.

Santos-Reyes, Jaime, Tatiana Gouzeva, and Galdino Santos-Reyes. "Earthquake Risk Perception and Mexico City's Public Safety." *Procedia Engineering* 84 (2014): 662–71.

Saramago, Victoria. *Fictional Environments: Mimesis, Deforestation, and Development in Latin America*. Evanston, IL: Northwestern University Press, 2020.

———. "Magueys and Machines: Narratives of Environmental Change in Mid-Twentieth-Century Mexico," *FORMA* 2, no. 1 (2023): 85–108.

Sbrizzi, Paul. "A Conversation with Everardo González (*Drought*)." *Hammer to Nail*, August 24, 2012. https://www.hammertonail.com/interviews/a-conversation-with-everardo-gonzalez-drought.

Schipani, Sam. "How to Save the Paradoxical Axolotl." *Smithsonian Magazine*, Jan. 8, 2018. https://www.smithsonianmag.com/science-nature/saving-paradoxical-axolotl-180967734.

Schiwy, Freya. *The Open Invitation: Activist Video, Mexico, and the Politics of Affect*. Pittsburgh, PA: University of Pittsburgh Press, 2019.

Sedgwick, Eve Kosofsky. *Touching Feeling: Affect, Pedagogy, Performativity*. Durham, NC: Duke University Press, 2003.

Segato, Rita Laura. "A Manifesto in Four Themes." Translated by Ramsey McGlazer. *Critical Times* 1, no. 1 (2018): 198–211.

Serafini, Paula. *Creating Worlds Otherwise: Art, Collective Action, and (Post)extractivism*. Nashville, TN: Vanderbilt University Press, 2022.

Serratos, Francisco. *El capitaloceno: Una historia radical de la crisis climática*. Mexico: Festina, 2020.

Seymour, Nicole. *Bad Environmentalism: Irony and Irreverence in the Ecological Age*. Minneapolis: University of Minnesota Press, 2018.

Shotwell, Alexis. *Against Purity: Living Ethically in Compromised Times.* Minneapolis: University of Minnesota Press, 2016.

Sipola, Simo, dir. *Murha Meksikossa.* Helsinki: Yleisradio, 2016.

Sisk, Christina L. *Mexico, Nation in Transit: Contemporary Representations of Mexican Migration to the United States.* Tucson: University of Arizona Press, 2011.

"Six Mexican Artists Revisit José Juan Tablada and His New York Circle." Proxyco Gallery (website), November 15, 2017. https://www.proxycogallery.com/talon-rouge.

Sluis, Ageeth. *Deco Body, Deco City: Female Spectacle and Modernity in Mexico City, 1900–1939.* Lincoln: University of Nebraska Press, 2016.

Smaill, Belinda. *Regarding Life: Animals and the Documentary Moving Image.* Albany: SUNY Press, 2016.

Smith, Amanda M. *Mapping the Amazon: Literary Geography after the Rubber Boom.* Liverpool, UK: Liverpool University Press, 2021.

Soler Frost, Pablo. *Oriente de los insectos mexicanos.* Oaxaca: Zopilote Rey, 2019.

Sontag, Susan. "Posters: Advertisement, Art, Political Artifact, Commodity." In *The Art of Revolution: Castro's Cuba: 1959–1970,* edited by Dugald Stermer. New York: McGraw-Hill, 1970.

Sosa-Nishizaki, Oscar, Felipe Galván-Magaña, Shawn E. Larson, and Dayv Lowry. "Conclusions: Do We Eat Them or Watch Them, or Both?: Challenges for Conservation of Sharks in Mexico and the NEP." *Advances in Marine Biology* 85 (2020): 93–102.

Soto-Coloballes, Natalia. "The Development of Air Pollution in Mexico City." *Journal of Environment and Development* 10, no. 2 (2020): 1–10.

Soto Curiel, Jesús Adolfo. *Recordar en presente: Cine documental y memoria en México.* Mexicali: Universidad Autónoma de Baja California, 2017.

Speranza, Graciela. *Lo que no vemos, lo que el arte ve.* Buenos Aires: Anagrama, 2022.

Spires, Adam. "Homero Aridjis and Mexico's Eco-Critical Dystopia." In *Blast, Corrupt, Dismantle, Erase: Contemporary North American Dystopian Literature,* edited by Brett J Grubisic, Gisele M. Baxter, and Tara Lee, 339–54. Waterloo, Canada: Wilfred Laurier University Press, 2014.

Starr, Douglas. "Just 90 Companies Are to Blame for Most Climate Change, This 'Carbon Accountant' Says." *Science,* August 25, 2016. https://www.science.org/content/article/just-90-companies-are-blame-most-climate-change-carbon-accountant-says.

Studnicki-Gizbert, Daviken. "Exhausting the Sierra Madre: Mining Ecologies in Mexico over the Longue Durée." In *Mining North America: An Environmental History since 1522,* edited by J. R. McNeill and George Vrtis, 19–46. Berkeley: University of California Press 2017.

Svampa, Maristella. *Neo-Extractivism in Latin America: Socio-Environmental Conflicts, the Territorial Turn, and New Political Narratives.* Cambridge, UK:

Cambridge University Press, 2019.

Tablada, José Juan. *Un día . . . (poemas sintéticos)*. Caracas: [Imprenta Bolívar], 1919. Available online at "Juan José Tablada: Vida, letre e imagen." UNAM, 2021. https://www.iifl.unam.mx/tablada.

Tenorio-Trillo, Mauricio. *I Speak of the City: Mexico City at the Turn of the Twentieth Century*. Chicago: University of Chicago Press, 2012.

Tetreault, Darcy. "Free-Market Mining in Mexico." *Critical Sociology* 42, no. 4–5 (July 2016): 643–59.

———. "Water in Zacatecas: A Crisis without Conflict." In *Social Environmental Conflicts in Mexico: Resistance to Dispossession and Alternatives from Below*, edited by Darcy Tetreault, Cindy McCulligh, and Carlos Lucio, 183–217. New York: Palgrave Macmillan, 2018.

———. "The New Extractivism in Mexico: Rent Redistribution and Resistance to Mining and Petroleum Activities." *World Development* 126 (2020): 1–10.

———. "Extractive Policies in Mexico at the Outset of López Obrador's Presidency." In *Latin American Extractivism: Dependency, Resource Nationalism, and Resistance in Broad Perspective*, edited by Steve Ellner, 149–66. Lanham, MD: Rowman and Littlefield, 2020.

Tetreault, Darcy, and Cindy McCulligh. "Water Grabbing via Institutionalized Corruption in Zacatecas, Mexico." *Water Alternatives* 11, no. 3 (2018): 572–91.

Thomas, Darren, Terry Mitchell, and Courtney Arseneau. "Re-evaluating Resilience: From Individual Vulnerabilities to the Strength of Cultures and Collectivities among Indigenous Communities," *Resilience* 4, no. 2 (2016): 116–29.

Tinoco, Óscar. "'Para mí el arte siempre va a tener que ver con la política': Betzabé García." Crash, May 16, 2016. http://www.crash.mx/encuadre/para-mi-el-arte-arte-siempre-va-a-tener-que-ver-con-la-politica-betzabe-garcia.

Toledo, Víctor M., David Garrido, and Narciso Barrera-Bassols. "The Struggle for Life: Socio-environmental Conflicts in Mexico." Translated by Mariana Ortega Breña. *Latin American Perspectives* 42, no. 5 (June 2015): 133–47.

Toledo, Víctor M. *Ecocidio en México: La batalla final es por la vida*. Mexico DF: Grijalbo, 2015.

Torres-Rodríguez, Laura J. *Orientaciones transpacíficas: La modernidad mexicana y el espectro de Asia*. Chapel Hill: University of North Carolina Press, 2019.

Torres Méndez, Martha. "Desert Landscapes: Violence and Enduring Subjects in Contemporary Mexican Literary and Visual Culture." Ph.D. diss., UC Irvine, 2020.

Tsing, Anna Lowenhaupt. *The Mushroom at the End of the World: On the Possibility of Life in Capitalist Ruins*. Princeton, NJ: Princeton University Press, 2015.

Tuck, Eve. "Suspending Damage: A Letter to Communities." *Harvard Educational Review* 79, no. 3 (2009): 409–428.

Uc Be, Pedro. *Resistencia del territorio maya frente al despojo*. México: Centro de estudios para el cambio en el campo mexicano, 2021.

US Fish and Wildlife Service, "Service Finalizes Changes to Mexican Wolf Experimental Population Rule in Arizona and New Mexico." News release, January 12, 2015. http://www.fws.gov/southwest/es/mexicanwolf/pdf/NR_Mexican_Wolf_f10j_FINAL.pdf.

Valadez Rodríguez, Alfredo. "Mina zacatecano, convertida en un cementerio tóxico." *La Jornada*, August 23, 2010. https://www.jornada.com.mx/2010/08/23/estados/031n1est.

———. "Nuevo Mercurio, una mina de contaminación." *Zacatecas Online*, August 24, 2010. https://zacatecasonline.com.mx/index.php/noticias/municipios/6932-mina-nuevo-mercurio-un-cementerio-toxico.

Valencia, Sayak. *Gore Capitalism*. Translated by John Pluecker. Los Angeles: Semiotext(e), 2018.

Van Dooren, Thom. *Flight Ways: Life and Loss at the Edge of Extinction*. New York: Columbia University Press, 2014.

Vardy, Mark, and Mick Smith. "Resilience." *Environmental Humanities* 9, no. 1 (2017): 175–79.

Vaughan, Hunter. *Hollywood's Dirtiest Secret: The Hidden Environmental Costs of the Movies*. New York: Columbia University Press, 2019.

Velázquez, Perla. "Entrevista Verónica Gerber Bicecci, homenajes a Tablada y a Ampáro Dávila." Bitácora de vuelos, January 16, 2020. http://www.rdbitacoradevuelos.com.mx/2020/01/entrevista-veronica-gerber-bicecci.html.

Vera-Morales, Luis R. "Dumping in the International Backyard: Exportation of Hazardous Wastes to Mexico." *Tulane Environmental Law Journal* 7, no. 2 (1994): 353–88.

Vergara, Germán. *Fueling Mexico: Energy and the Environment*. Cambridge: Cambridge University Press, 2021.

"Verónica Gerber: *La máquina distópica*." Seminario Investigación Poéticas de lo Inquietante, January 25, 2021. https://www.youtube.com/watch?v=tZtjCLTodTo&t=5845s.

Villanueva, Paloma. "La mina que dividió a un pueblo." Oxfam México, May 10, 2018. https://www.oxfammexico.org/historias/la-mina-que-dividi%C3%B3-un-pueblo.

Villeda, Karen A. *Dodo*. México: Consejo Nacional para la Cultura y las Artes, 2013.

Viramontes, Sofía. "Betzabé García y el devenir en la reconstrucción." *Gatopardo*, March 31, 2017. https://gatopardo.com/arte-y-cultura/betzabe-garcia-reyes-pueblo-no-existe.

Vitz, Matthew. *A City on a Lake: Urban Political Ecology and the Growth of Mexico City*. Durham, NC: Duke University Press, 2018.

Volke Sepúlveda, Tania, and Jan Antonio Velasco Trejo. *Tecnologías de remediación para suelos contaminados*. Mexico: Institutio Nacional de Ecología, 2002.

"Voyager—What's on the Golden Record." NASA (website), accessed August 8, 2021. https://voyager.jpl.nasa.gov/golden-record/whats-on-the-record.

Walker, Janet, and Nicole Starosielski. "Introduction: Sustainable Media." In *Sustainable Media: Critical Approaches to Media and Environment*, edited by Nicole Starosielski and Janet Walker, 1–20. New York: Routledge, 2016.

Wang, Dorothy J. *Thinking Its Presence: Form, Race, and Subjectivity in Contemporary Asian American Poetry*. Stanford, CA: Stanford University Press, 2014.

Weisbrot, Mark, Stephan Lefebvre, and Joseph Sammut. "Did NAFTA Help Mexico?: An Assessment after 20 Years." *CEPR Reports and Issue Briefs*, Center for Economic and Policy Research, Feb. 2014. https://www.cepr.net/documents/nafta-20-years-2014-02.pdf.

Whyte, Kyle. "Indigenous Climate Change Studies: Indigenizing Futures, Decolonizing the Anthropocene." *English Language Notes* 55, no. 1–2 (Fall 2017): 153–62.

"Widespread Drought in Mexico." NASA Earth Observatory. NASA (website), accessed February 2, 2022. https://earthobservatory.nasa.gov/images/148270/widespread-drought-in-mexico.

Williams, Raymond. *Marxism and Literature*. Oxford, UK: Oxford University Press, 1977.

Worley, Paul M., and Rita M. Palacios. *Unwriting Maya Literature: Ts'íib as Recorded Knowledge*. Tucson: University of Arizona Press, 2019.

Wright, Melissa W. "Visualizing a Country without a Future: Posters for Ayotzinapa, Mexico and Struggles against State Terror." *Geoforum* 102 (2019): 235–41.

Wylie, Lesley. *Colonial Tropes and Postcolonial Tricks: Rewriting the Tropics in the novela de la selva*. Liverpool, UK: Liverpool University Press, 2009.

Yusoff, Kathryn. *A Billion Black Anthropocenes or None*. Minneapolis: University of Minnesota Press, 2018.

Zapata, Isabel. *Una ballena es un país*. Mexico DF: Almadía Ediciones, 2019.

Zavala, Lauro. "El nuevo documental mexicano y las fronteras de la representación." *Toma Uno* 1 (2012): 25–36.

Zavala, Oswaldo. *Los cárteles no existen: Narcotráfico y cultura en México*. Barcelona: Malpaso, 2018.

Zelizer, Barbie. *About to Die: How News Images Move the Public*. Oxford, UK: Oxford University Press, 2010.

Zepeda, Eduardo, Timothy A. Wise, and Kevin P. Gallagher. "Rethinking Trade Policy for Development: Lessons from Mexico under NAFTA." *Carnegie Policy Outlook*, Dec. 2009. https://carnegieendowment.org/2009/12/07/rethinking-trade-policy-for-development-lessons-from-mexico-under-nafta-pub-24271.

Index

Page numbers in *italic* refer to figures.

accumulation, 6, 79, 113, 166
    of capital, 10–11
    capitalist, 116, 175
    colonial-capitalist, 40
    by dispossession, 3
    settler-colonial, 20
    state extractivism and, 222n43
activism, 20, 86, 205n56
    Aridjis and, 96
    documentary and, 121
    ecofilm and, 126
    environmental, 98–99, 152
    Indigenous, 79
    Mexican poetry and, 98–99
aesthetics, 3, 126, 137, 141
    cinematic, 172
    of counterfactual mourning, 59, 89
    forensic, 3–4, 176
    landscape, 26–27, 55
    Mexican, 31, 33
    nationalist, 34
    nonhuman, 35
    queer, 79
    of remembrance, 71
    rural, 123
    of *Sembremos suenos y cosechemos*
        esperanzas (Lapiztola), 77
    small-file, 159
    territory and, 175
    violence against animals and, 103
affect, 5, 125
agency, 64
    cultural, 20
    Dutch imperial expansionism and, 101
    female, 46
    individual, 55, 177
    migration and, 120, 214n12
    narrative, 142
    political, 15, 70
agriculture, 14, 117, 119, 175
Aguilar Gil, Yásnaya Elena, 65, 149, 154–55
Álvarez Franco, Mónica, 171–72, 178
Andermann, Jens, 123, 127, 176
animals, 34, 91, 114, 126, 128
    domesticated, 112–13
    endangered, 16, 100
    environmental crisis and, 208n17
    farm, 130
    in literature, 115
    violence against, 103
    See also bestiary; *specific species*

Anthropocene, 6, 18, 20, 25, 54–55, 189n60, 189n63, 190n1
   Benjamin and, 192n17
   rewriting and, 27–28
anthropocentrism, 93, 114, 116, 178
Aquí Cine, 167–68, 173
Aridjis, Homero, 14–15, 95–97, 209n20, 209n22
Armas, Marcela, 10, *11*
art, 2, 4–7, 19, 53–55, 86, 89–90, 151, 175, 177–81
   antiextractivist, 16, 79, 199n91
   commons and, 23, 30
   environmentalist, 3, 13, 79
   evidentiary function of, 6
   extinction and, 96, 116
   extractivism and, 20
   Mexican, 8, 31, 192n28, 199n91
   proximity and, 94
   public, 10, 77–78, 179 (*see also* Lapiztola)
   street, 73, 77
   territory and, 22
Asamblea Nacional de Afectados Ambientales (ANAA, National Assembly of Affected Environments), 15, 202n34
as if, the, 4, 28, 54–56
authorial control, 26–28
autonomy, 73
   Indigenous, 8, 15, 72, 98
   territorial, 15
   Triqui, 77
axolotl, 80–81, 92. *See also* Rincón Gallardo, Naomi: *viaje de formol, El*
Ayotzinapa massacre, 3, 59, 64–65, 71–72, 89, 98–99, 205n56

Baja Studios, 169–70, 173
Banco de México, 16, 91
Barthes, Roland, 29, 35, 55, 118

Bashō, 33, 35
Beckman, Ericka, 10, 185n31
Benjamin, Walter, 30, 40, 54, 82, 191n15, 192n17, 199n87
bestiary, 33, 193n32
bicycles, 22, 162–65, 166
biodiversity, 7, 91, 92–93, 95
body-territory, 59, 78–79

Calakmul Biosphere Reserve, 12, 97, 209n25
Calderón, Felipe, 10, 17, 41
cambio, El (Joskowicz), 59–62, 64–65
cantos cardenches, 127, 216n37
capital, 8, 10, 62–63, 67, 99, 110, 145–46, 155
   accumulation of, 11
   cinema and, 152
   cultural infrastructures and, 164
   ecocidal, 12
   extractivist, 199n87
   global, 157, 176
   grammar of, 102
   infrastructure and, 154
   language of, 109
   political, 95
   resilience and, 120
   sublime and, 35
   transnational, 62
   victims and, 74
capitalism, 38, 51, 62, 68, 82, 131, 175
   colonial, 18, 173
   communicative, 4
   ecohorror and, 198n79
   extractive, 35, 56
   extractivist, 8–9, 19, 22, 28, 89, 95, 180
   gore, 128
   green, 163
   modernity and, 166
   neoliberal, 114
   the rural and, 144

Cariño, Alberta "Bety," 74–75, 77–79, 86, 88, 205n58. *See also* Lapiztola; Rincón Gallardo, Naomi: *trilogía de cuevas, Una*; Rincón Gallardo, Naomi: *viaje de formol, El*
catastrophe, 59
　ecological, 1–2, 183n8
　environmental, 137
Centro de Apoyo Comunitario Trabajando Unidos (CACTUS), 74, 78
Centro Mexicano de Derecho Ambiental (CEMDA, Mexican Center for Environmental Law), 63, 201n24
Cindy la Regia (Aguilar Mastretta and Limón), 171, 173
Cine en tu Comunidad, El, 161, 163
cinema, 157, 131, 141, 155, 158, 170
　access to, 163
　ambulatory, 22
　art house, 124
　consumption of, 150
　contemporary, 133
　emissions and, 166, 168, 172
　environmental, 124, 149, 152
　extractivism and, 178
　green, 171
　infrastructures of, 151, 168
　Latin American, 123, 153
　of migration, 120, 145, 214n12
　oil and, 152, 156, 160, 172
　postcarbon, 152–53, 168
　rural Mexico and, 22, 132
　slow, 127
　spectatorship, 173
　See also Mexican cinema
Cine Móvil ToTo, 22, 153, 160–68, 173, 178. *See also* bicycles
Coahuila, 117, 122–23, 126, 131, 149
Coca-Cola, 16, 179–80, 195n60
collaboration, 31, 55, 166

coloniality, 8, 152, 177
commodification, 13, 17, 95, 104, 203n42
　of commons, 78
　of labor power, 102
　of life, 22, 105, 108
　of suffering, 125
　of territory, 63
commons, 30, 78, 151, 172, 183n2
　technology and, 222n43
　territorial, 23, 65
　text as, 36
community, 59, 65–66, 128, 173, 183n2
　benefits of, 134
　dispossession and, 1
　maintenance, 135
　making, 127
　modernity and, 166
compañía, La (Gerber Bicecci), 21, 28, 30–31, 39–44, 47–48, 49–54, 177–81
　class dynamics in, 199n92, 199n87
　as ecohorror, 194n47
　FEMSA and, 196n60
　geological writings and, 191n11
　tequio and, 196n62
　See also Dávila, Amparo: "huésped, El"; rewriting
comuneros, 133–34, 147
conservation, 95–96
　efforts, 91, 99
　rhetoric, 113
conservationism, 15, 113
contamination, 50, 198n78
　air, 12, 54
　industrial, 61
　PCB, 49
　risk, 57
　water, 12, 180, 184n8, 196n60, 197n73
contiguity, 8, 21–22, 50, 94–95, 100, 110, 116, 177
　interspecies, 93

contingency, 3, 5, 19, 32, 35, 55, 181
of water, 145
Coronil, Fernando, 17, 152
Corriente del Golfo, La, 149–50. *See also El tema*
costumbrismo, 123, 196n65
COVID-19 pandemic, 92, 150, 162, 169, 201n26
Coyolxauhqui, 83–87. *See also* Rincón Gallardo, Naomi: *El viaje de formol*
Cuates de Australia (González), 22, 117, 122–23, 125–32, 141–42, 144–47, 157, 217n46
sensory ethnography and, 134
superstition and, 216n44
See also drought
cultural infrastructure, 151, 160, 164, 166–67, 170, 178
erosion of, 8, 175
expansion in, 15

Dávila, Amparo, 21, 28, 31, 54–55, 196n65, 199n87
"huésped, El," 40, 42, 44–48, 50–53, 199n92
See also compañía, La (Gerber Bececci)
deforestation, 7, 9, 12, 41, 117, 142, 145
drought and, 209n25
Pemex and, 186n41
de la Cadena, Marisol, 4, 203n37
del Angel, Elizabeth, 46, *47*
dependency, 1, 5, 6
deregulation of film, 150, 156, 201n18
Derrida, Jacques, 115–16, 200n8
desiccation, 21, 25, 54, 128–29, 190n1
desire, 2, 4, 6, 23, 51
counterfactual mourning and, 59, 80
subjunctive and, 184n15
development, 12–13, 15, 40, 88, 117, 160
conflicts, 137
downsides of, 96

extractivist, 7, 10, 17, 177
extreme, 170
favorable, 60
national, 64, 72, 155, 214–15n19
real estate, 111
state-sponsored, 136
sustainable, 16, 163
in Xochimilco, 92
Díaz, Floriberto, 65–66, 196n62
Díaz, Porfirio, 30, 192n27
Dionicio, Rosalinda, 69, 81, 88. *See also* Rincón Gallardo, Naomi: *Resiliencia tlacuache*
disappearance, 3, 31, 63–65, 202n32
dispossession, 2–3, 6, 77, 137–38
extractivism and, 1–2, 58–59, 79
Indigenous, 1, 78
racialized and classed dynamics of, 147
resilience to, 140
in *El sueno de toda célula* (Guerrero), 111
in *Una trilogía de cuevas* (Rincón Gallardo), 80, 83
dodo bird, 29, 99–101, 116. *See also* Villeda, Karen: *Dodo*
doubt, 4–6, 19, 36, 152, 176–77
drought, 6, 8, 119, 122–24, 130, 164, 192n17
fracking and, 186n39
in Mexico, 14, 18, 209n25, 213n9
narrative of, 126
See also Cuates de Australia (González)

Echeverría, Bolívar, 11–12, 166, 175, 194n51
ecocinema, 22, 121, 124–25, 147, 150, 173, 219n7
Latin American, 152
Mexican, 15, 157
ecocultural production, 8, 13–15, 176

ecodocumentaries/ecodocs, 119, 121, 124–26, 133, 149, 157–58, 214–15n19
ecological collapse, 18–19, 39
ecosystems, 91, 176, 221n43
　collapsing, 121
　elimination of, 186n41
　extractivist, 62
ejidos, 57, 122, 126–28, 131
emissions, 152, 158
　carbon, 9, 138, 168–70
　global, 3, 9, 159, 166
　greenhouse gas, 7, 12, 159, 178
　mining and, 36, 195n57
　reduction of, 16–17, 154, 166–69, 172
　tracking of, 171
　See also Pemex (Petróleos Mexicanos)
Emmelhainz, Irmgard, 7, 89, 99
endangered species, 14, 16, 96. *See also* animals: endangered
environmental crisis, 1–4, 8–10, 14, 16, 126, 145–46, 149
　animals and, 208n17
　cinema and, 152
　developmentalism and, 157
　documentary and, 125
　images of, 118
　manufactured, 137
　Mexican literature about, 27
　national project and, 54
　rewriting and, 30
environmental damage, 4, 16–17, 120, 147, 195n60
environmentalism, 7, 65
　first-wave, 60
　land defense and, 66, 79, 187n45
　Mexican, 13, 15–16, 61, 187n45
　poetry and, 97
　PVEM and, 188n54
environmentalist art, 3, 13, 79
exoticism, 33, 140

extinction, 2, 21, 93–97, 103–4, 110, 209n20
　of dodo bird, 100–101
　human, 36, 38, 56
　insects and, 13
　Mexican gray wolf and, 113–14
　poetry, 99, 116
　settler, 102
　shark finning and, 106–7
　sixth, 8, 91, 111, 116, 175
　supplement and, 116
　of Tasmanian tiger, 92
　Tren Maya project and, 98
　See also under poetics
extractivism, 2, 7–8, 12, 22–23, 48, 72, 195n53
　art and, 20, 180–81
　capitalist, 38, 175, 177
　cinema and, 152, 155, 178
　colonial model of, 9
　in *La compañía* (Gerber Bececci), 42–43, 52, 56, 177, 180–81, 196n60
　counterfactual mourning and, 59, 90
　crisis of, 175
　critiques of, 151, 178
　cultural, 66, 203n42
　developmentalist faith in, 45
　dissent to, 58
　distance and, 52–53
　exchange rate of, 67
　FEMSA Biennial and, 195–96n60
　green, 221n43
　inequalities of, 49
　land defense and, 61
　Mexico's petronationalism and, 10
　public arts funding and, 179
　public space and, 69–70
　recompensatory model of, 68
　in San Felipe Nuevo Mercurio, 40
　simultaneities of, 50–51
　social programming and, 160
　state, 222n43
　toxic masculinity and, 193n47

extractivism (*continued*)
  trilogía de cuevas, Una (Rincón Gallardo) and, 79, 84
  violence and, 61–62, 74, 89, 201n18
  in Zacatecas, 41

Falicov, Tamara, 123, 136, 163
feminicide, 64, 72
FEMSA, 16, 56, 179–80, 195–96n60
  Biennial, 16, 32, 41–43, 53, 178–80, 195n60
  foundation, 178–79, 195n60
  See also compañía, La (Gerber Bicecci); greenwashing
fiction, 5, 176
  consent as, 58
  Dávila's, 45
  environmentalist, 194
  fantastical, 44
  film, 121
  genre, 196n65
  poetry and, 208n8
  postapocalyptic, 138
  speculative, 5, 119, 140 (*see also* speculation)
  young adult, 13
film, 7–8, 119, 177
  consumption in Mexico, 150
  distribution, 122, 140, 150, 152, 158, 161
  documentary, 99, 118
  equipment, 121
  exhibition, 150–52, 157–58, 161–62, 168
  industry, 151, 156–57, 168–72
  Latin American, 123
  production, 152, 168–69, 171
  projection, 166
  technologies of, 153
  viewers, 163
film festivals, 69, 118, 137, 140, 168
  environmentalist, 15, 122, 157, 214n19, 221n35
  greening at, 171

flooding, 16, 18, 119, 133, 144
  mangrove swamps and, 154
  in Mexico City, 190n1
  in *El Remolino* (Herrero Garvín), 142, 146
  in *Los reyes del pueblo que no existe* (García), 136
Flores Soberanes, Samir, 57–58, 63, 90, 189n58, 200n5
foreclosure, 2, 89, 175
  ecocidal, 21
  environmental, 20
  extinction and, 94, 96
  extractivist capitalism and, 8, 22
  of future, 6
  of life, 4
  of political dissent, 58
  regional, 154
  rural Mexico and, 117
  Zacatecas and, 51, 53
Fortuna Silver Mines, 69, 181. *See also* Minera Cuzcatlán
fossil fuels, 9, 53, 155, 159–60
  alternatives to, 165
  infrastructure and, 120, 154
  reliance on, 172–73
Fuentes, Carlos, 15, 25, 196n65
futurity, 75, 177
  bicycle and, 165
  of collective creative practices, 84
  crisis of, 14
  of land defense, 59
  without optimism, 146
  of plastic, 38
  of postextractivist forms, 89
  rural, 131
  Zapotec, 6

García, Betzabé
  reyes del pueblo que no existe, Los, 22, 117, 133–42, 144–47, 151, 178
  Unsilenced, 134, 137, 140
  Venecia, Sinaloa, 136–37, 140

García Bernal, Gael, 149, 220n32. *See also tema, El*
García Hernández, Álvaro, 40, 199n85
Gerber Bicecci, Verónica, 8, 43, 53–56, 177–78, 195n59
  antiextractivist art of, 16
  Conjunto vacío, 31
  ecohorror and, 198n79
  future tense and, 50–51
  máquina distópica, La, 42, 180, 196n60
  Otro día . . . (poemas sintéticos), 21, 28, 30–32, 34–39, 54 (*see also* Tablada, José Juan)
  tierra es plana como una hoja (y cabalga en el aire), La, 32
  See also compañía, La; Dávila, Amparo; rewriting
Ghosh, Amitav, 4–5, 18, 54
Global North, 9, 18, 123
Global South, 9, 18, 123, 153, 159
Golden Record, 36–38
Gómez-Barris, Macarena, 7, 48, 176
González, Everardo, 122–23, 125–29, 132, 137. *See also* Cuates de Australia
González de Arce, Rocío, 121, 214n17
greenwashing, 16, 164, 179, 196n60
Grupo de los Cien, 15, 96, 98
Grupo Modelo, 164, 179
Guerrero, Maricela, 8, 45, 93, 116
  sueño de toda célula, El, 22, 95, 110–15, 177
  See also poetics: extinction

haiku, 32–36, 39, 193n32. *See also* Gerber Bicecci, Verónica: *Otro día . . . (poemas sintéticos)*; Tablada, José Juan: *Un día . . . (poemas sintéticos)*
Heise, Ursula, 95, 103
Herrero Garvín, Laura, 142, 216n36
  Remolino, El, 22, 117, 125, 142–46, 157, 216n36

Higashi, Alejandro, 98, 103
Huachicolero (Nito), 61, 121
Humboldt, Alexander von, 18, 25, 27, 54, 83

identity, 77, 88, 91
  Mexican, 25, 54, 92
impunity, 72–74, 77–78, 89–90, 128, 145
  aesthetics of remembrance and, 71
  state, 3, 63
  violence against land defenders and, 82, 201n18
incendio de la mina El Bordo, El (Herrera), 3–4, 29
indicative, the, 2–3
indigeneity, 67, 73
Indigenous culture, 66, 77
Indigenous politics, 65–66
infrastructure, 127, 133, 152, 157–58, 178, 188n58, 223
  of cinema/film, 151, 153, 156, 168, 170
  of cultural production, 151
  of distribution, 150, 167, 173
  exhibition, 22, 150, 157–58, 160–61, 166–68, 220n32
  extractivist, 181
  formal, 28, 31
  fossil fuel, 120, 153–55, 172
  institutional, 15
  megaprojects, 134
  neo-extractivism and, 195n53
  of Oaxacan culture, 69
  state, 179
  state-funded, 137
  textual, 44, 47
  of violence, 89
  See also cultural infrastructure
Instituto Mexicano de Cinematografía (IMCINE, Mexican Film Institute), 157, 161–63
internet, 35, 128, 150, 158
intertextuality, 29–30

Jaakkola, Jyri Antero, 74, 78
jaguar, 91, 97–98
Japan, 33–34
justice, 65, 78, 89–90, 134, 154, 180
  energy, 168
  environmental, 141, 150, 170, 172–73
  social, 72, 159, 173, 204n53
  system, 59, 90

labor, 9, 17–18, 75, 110, 168–70, 175, 194n52, 211n42
  changes, 213n9
  conditions, 40
  disputes12;
  I-machinarius (Armas) and, 11
  power, 102, 201
  relations, 172
  rural, 128
Laguna Ojo de Liebre, 95–96
Laguna San Ignacio, 95–96, 209n20
land, 1, 64, 72–73, 103, 181, 205n56
  arable, 143
  capitalist understandings of, 65
  communal, 126, 203–4n48
  community and, 66, 183n2
  degradation of, 186n41
  disputes, 58, 64, 164
  exploitation of, 175
  expropriation of, 9, 133
  extractivism and, 45, 83, 89
  Indigenous, 74
  mining and, 62
  pollution of, 41
  redistribution of, 20
  safe access to, 49–50
  Zoque, 98
land defenders, 8, 10, 88, 90, 175, 201–2n26, 214–15n19
  forced disappearance of, 202n32
  Indigenous, 17, 21, 65, 67
  murder of, 14, 58–64, 68, 71–74, 78, 80, 82, 89, 177, 201n18, 201n24
  public homages to, 77, 81

land defense, 6–7, 16, 59–61, 64–65, 72, 78, 86, 96, 121
  CEDMA and, 201n24
  Indigenous women's leadership in, 79
  invisibility of, 187n45
  pluriversal, 177
  in *Una trilogía de cuevas* (Rincón Guerrero), 80, 82, 89
  See also territory: defense of
Lapiztola, 75–77
Latin America, 18–19, 71
  cinema in, 153, 156, 221n35
  consent to extractive projects in, 58
  El Dorado myth and, 185n31
  environmental cinema in, 152
  environmental movements in, 65, 214n19
  extractivism and, 160
  film festivals in, 122
  Latin American Water Funds Partnership and, 179
  mining in, 10, 197n73
  violence against land defenders in, 62
Levinas, Emmanuel, 70, 94
literature, 3, 7, 19, 94, 99, 151, 180
  Amazonian, 203n42
  axolotl in, 92
  children's, 100
  climate crisis and, 55
  extinction and, 116
  Latin American, 4
  Mexican, 8, 21, 27–28, 33, 183n8, 191n14, 196n65
  about Mexico City, 25
  of nineteenth century, 131
  rewriting and, 29
  subjunctive capacity of, 115
  See also poetry
lithium, 166, 221n40, 221–22n43
Lomnitz, Claudio, 10, 20
López Obrador, Andrés Manuel (AMLO), 6–7, 17, 189n58, 194n53, 202n28

jaguar and, 97–98
  lithium mining and, 221n40, 222n43
  Tren Maya and, 209n25
López Portillo, José, 95, 156
Luna, Diego, 149, 220n32

maize, 73–74, 81
Marán, Luna, 167, 181
  "Piel territorio," 1, 3, 5–7
  Tío Yim, 162
Marcone, Jorge, 124, 214n19
Martínez Pinacho, Tomás, 70–72
masculinity, 44, 103, 193n47
  Mexican, 128, 216n44
Mazatlán, 133, 138, 151, 164
Melchor, Fernanda, 197n65
  Temporada de huracanes, 62
mercury, 40, 41, 49
Mexican cinema, 147, 150–51, 157, 168, 171
  access to, 153, 161, 165
  circulation of, 22
  masculinity in, 216n44
  oil and, 155–56
  postcarbon cultural production and, 173
  provincia and, 117, 132
  the rural and, 145
Mexican gray wolf, 91, 111, 113
Mexican Revolution, 25, 33, 72, 131–32, 187n45
  Rulfo's critique of, 133
  Tablada and, 192n27
Mexico City, 25, 70, 84, 132, 144
  air pollution in, 26, 54, 60, 96
  Chapultepec Zoo, 16
  ecodocs and, 213n8
  exhibition infrastructure and, 157
  Gerber Bicecci and, 43, 53
  urban ecocide and, 15
  water management in, 190n1
migration, 8, 10, 40, 125, 131, 213n9
  cinema of, 120, 145, 214n12

climate, 120, 127
  forced, 128
  narratives, 145–46
  rates, 119
  rural-to-urban, 14, 117
Minera Cuzcatlán, 69–70, 78, 81, 88
  sponsorship of Oaxaca FilmFest, 69, 181, 203n47
mining, 1, 48, 98, 121
  disasters, 3, 29
  foreign direct investment in, 10
  illegal, 202n26
  laws, 62
  lithium, 221n40, 221n43
  in Mexico, 50–51, 195n57, 197n73
  open-pit, 17
  sand, 63
  transnational, 8
  in Zacatecas, 40–42, 45, 53, 80, 180, 194n48, 194nn52–53 (*see also* Dávila, Amparo; Rincón Gallardo, Naomi: *Sangre pesada*)
  See also compañía, La (Gerber Bicecci); Fortuna Silver Mines; mercury; *Noche de fuego* (Huezo); silver
Mining Law (1892), 40, 41
modernity, 2, 10–12, 18, 60, 117, 163, 166
  carbon-intensive, 6
  cinema and, 160
  European, 103
  extractivist, 11, 59
  fossil-fueled, 9, 19, 152
  Japan and, 33
  media and, 153
  Mexican, 95
  nature and, 35
monarch butterfly, 91, 96, 209n22
Monsanto, 12, 49
Monsiváis, Carlos, 13–14, 131–32
montage, 43, 51
Moore, Jason W., 95, 189n60

mourning, 58, 64–65, 72
  counterfactual, 8, 21, 58–59, 68, 78–79, 89, 94, 177
  irreverent modes of, 80
  public, 69, 77, 90
Museo de Arte Abstracto Manuel Felguérez, 31–32, *42*, 43

NAFTA, 17–18, 119, 145, 191n14, 197–98n74
  film distribution and, 150, 156
  migration and, 213n9
nationalism
  green corporate, 164
  nostalgic, 73, 86
  petronationalism, 10
  resource, 9
NGOs, 48, 78
#NiUnaMenos, 64, 72, 98–99
Noche de fuego (Huezo), 61, 121

Oaxaca (state), 63–64
  activists in, 181
  energy projects in, 202n28
  festival dance in, 83
  Indigenous communities in, 164, 167
  as mine, 69–70, 74, 88–89
  seizure of Indigenous lands in, 203–4n48
  See also Minera Cuzcatlán
Oaxaca City, 75, 77–78
oil, 10–11, 96, 120, 151, 154, 156, 179
  cinema/film and, 152–53, 155, 160, 172
  crash, 62
  pipelines, 157
  resource nationalism and, 9
  See also Pemex (Petróleos Mexicanos)
ore, 41, 46, *47*
organized crime, 128, 134, 146, 197n73

Pacheco, José Emilio, 14, 95–97, 208n17
Paley, Dawn, 62, 64, 201n15

Partido Revolucionario Institucional (PRI, Institutional Revolutionary Party), 6, 61, 86
Partido Verde Ecologista de México (PVEM, Ecological Green Party of Mexico), 15–16, 188n54
peasantry, 73, 132–33
Pemex (Petróleos Mexicanos), 7, 10, 154–55, 186n41
  emissions and, 9, 12
  privatization of, 186n32
Peña Nieto, Enrique, 6–7, 17, 41, 194n53
Picachos Dam, 133–34, 136, 138, 140, 151, 178
pipelines, 57, 157, 200n1
poetics
  of Aridjis, 97
  environmental, 37
  extinction, 93–95, 99, 116, 177
  Guerrero's, 177
  of haiku, 35
  Mexican, 34
  modernista, 33
  of *Otro día . . . (poemas sintéticos)* (Gerber Bicecci), 37
poetry, 6, 29, 34, 38, 93–95, 99, 111–12, 115
  activism and, 98–99
  capaciousness of, 208n8
  Cariño's, 78
  experimental, 32
  extinction, 96–97, 99, 116
  NAFTA as, 191n14
  proximity and, 106, 110
  as resistance, 84
  See also Aridjis, Homero; Gerber Bicecci, Verónica; Guerrero, Maricela; Pacheco, José Emilio; Rodríguez Mendoza, Xitlálitl; Tablada, José Juan; Villeda, Karen; Zapata, Isabel
pollution, 14, 92, 170
  air, 26, 96, 190n1
  carbon, 9

light, 34
Pemex and, 186n41
polychlorinated biphenyl (PCB), 49–50
Potosí, 40, 45
praxis, 152, 178
  revisionary, 31, 39, 53–54
  of rewriting, 177
privatization, 62, 156, 163
proximity, 22, 93–94, 105–6, 110, 147, 177
PROXYCO, 31, 34
Proyecto Integral Morelos, 57, 70

Quintana-Navarrete, Jorge, 14, 201n21

racism, 50, 65, 189n63, 197–98n74, 202n32
representation, 3, 21, 92–94, 110, 152
  dodo bird and, 101
  enjambment and, 105
  rural, 157
  as supplement, 115–16
resilience, 81, 127, 130–31, 136, 144, 146–47, 178
  environmental crisis and, 157
  magical, 140, 147
  rural, 22, 118, 120, 125–26
  See also rural resilience film
rewriting, 8, 25, 94, 191n14, 199, 207
  environmental, 21, 27–30, 32, 54–56
  Gerber Bicecci and, 31, 34–41, 44, 50, 53–56, 177, 181, 195n59, 199n87, 199n92
  planetary, 27, 53, 55
  Tablada and, 33–34
Reyes, Alfonso, 14, 21, 55
  "Palinodia del polvo," 25–28, 39, 54, 194n47
  Visión de Anáhuac (1519), 25–26, 54, 190nn1–2
Rincón Gallardo, Naomi, 16, 42, 177–78
  Resiliencia tlacuache, 69, 74, 79–81, 83
  Sangre pesada, 79–80, 178–80

trilogía de cuevas, Una, 21, 59, 79–80, 82–83, 87–90
viaje de formol, El (*The Formaldehyde Trip*), 79–80, 83–88, 92, 206n62, 206n73
Rivera Garza, Cristina, 55, 62, 64, 69
  Autobiografía del algodón, 29
  cresta de Ilión, La, 44
  on geological writings, 28, 56, 191n11
  invencible verano de Liliana, El (*Liliana's Invincible Summer*), 183–84n8
Rodríguez Mendoza, Xitlálitl, 45, 93, 116
  Jaws, 22, 29, 95, 104–11
Rogers, Charlotte, 10, 176, 185n31
Rojas, Marisol, 77, 205n58
Rosas Mantecón, Ana, 157–58
Rosicler, 49–51, 180
Rulfo, Juan, 14, 45, 132–33, 196n65
runaway productions, 154, 169
rural, the, 61, 117–19, 131–32, 136, 140, 144; 145–46
rural resilience film, 117–18, 121, 124–26, 142, 145–48, 150, 178
  migration and, 120, 131
  rural Mexico and, 132
  See also García, Betzabé: *reyes del pueblo que no existe, Los*; González, Everardo; Herrero Garvín, Laura: *Remolino, El*

Sagan, Carl, 36–37
Sánchez, Mikeas, 67–68, 73, 98, 204n53
Sánchez Prado, Ignacio, 98, 124, 156
San Felipe Nuevo Mercurio, 40, 42–43, 46–51
  mine, 46, 47, 49
Saramago, Victoria, 14, 176
scarcity, 10, 109, 168, 173
  food, 125
  water, 22, 27, 57, 117, 145, 213n8
Secretariat of the Environment and Natural Resources' (SEMARNAT), 7, 10

sensorial immersion, 8, 118, 122, 126–27, 142, 147
sensorial/sensory ethnography, 117–18, 127, 134, 140, 142–43, 146, 151
Serrano, Roberto, 161–62
shark, 104, 111, 116
  finning, 105–7
  See also Rodríguez Mendoza, Xitlálitl: *Jaws*
silver, 40–41, 45, 194n48. *See also* Fortuna Silver Mines
Sinaloa, 117, 133, 136, 141. *See also* Mazatlán; Picachos Dam
Soler Frost, Pablo, 13–14
speculation, 5–6, 126, 175–77
  audience, 138
  colonial, 185n31
  subjunctive, 167
state, the, 22, 50, 85, 140, 147, 157
  Ayotzinapa massacre and, 71
  carbon pollution and, 9
  charro and, 128
  cinema and, 152
  clean energy and, 155
  climate change and, 16
  counterfactual mourning and, 59
  critiques of, 8
  cultural infrastructures and, 160, 163
  developmentalism and, 10, 15
  extractive economy and, 58
  fatalism and, 97–98
  Indigenous culture and, 66
  Indigenous past and, 92
  Indigenous peoples and, 210n25
  land defense and, 64
  Mexican gray wolf and, 112
  migration and, 131
  murder by, 77
  neo-extractivism and, 195n53
  racism and, 65
  resilience and, 120, 190n1
  San Felipe Nuevo Mercurio mine and, 49
  Tlatelolco massacre and, 61
  vandals and, 204n53
  violence and, 86
  wind energy and, 202n28
  writing and, 56
streaming, 158–59
  bandwidth, 173
  capabilities, 161
  content, 150
  services, 137
subjunctive (mood), 1–6, 19–20, 23, 28, 55, 114, 176–77
  Bello on, 184n15
  documentary and, 118
  dreaming in, 109
  mourning and, 58, 88, 94
  subordination and, 5, 32
  in *El sueño de toda célula* (Guerrero), 111, 114–15
  utopian potential of, 22
  writing and, 183n6
subjunctive voice, 119
  of visual, 118, 124, 134, 136, 138, 141, 147
subordination, 5–6, 19, 27, 32, 54, 177
sueño de Mara'akame, El (Cechetti), 121, 162
suicide, 5, 14, 54, 208n8
supplement, 115–16
Svampa, Maristella, 9, 52, 62, 195n53

Tablada, José Juan, 28, 31–32, 36, 44, 54–55, 192nn27–28
  Un día . . . (poemas sintéticos), 21, 32–35, 37–38, 193n32
tema, El, 149–50, 153–55, 157–58, 160, 173
Tenochtitlan, 19, 91–92, 190n1
territory, 1, 6–8, 12, 70–71, 81, 89
  acquisition of, 201n15
  art and, 22
  Chilean Anthropocene manifesto and, 20

commodification of, 17, 63
counterfactual mourning and, 59
defense of, 65, 80, 82, 90
ethical engagement with, 44
future/futurity of, 146, 177–78
Indigenous, 66–67
Indigenous approaches to, 183n2, 201n21
instrumentalist approaches to, 40
loss of, 133
of Mexican gray wolf, 111–13, 212n65
people and, 155
profit and, 21, 90, 177
relations/relationships with, 2, 4, 7, 20, 23, 58, 175
rewriting and, 177
subjunctive iterations of, 88
Zapotec, 82
See also body-territory
terror, 46, 52, 61, 69, 72, 82, 89
Tirado, Octavio Atilano Román, 133–34, 137
Tlatelolco massacre, 59–61, 187n45
Toledo, Víctor, 11, 65
Torres, Diego, 161–62
Tren Maya, 12, 97–99
Tuck, Eve, 51, 80

Uc Be, Pedro, 66, 98
urbanization, 25, 27, 92, 96

Valley of Mexico, 19, 21, 25, 111
Vásquez Sánchez, Bernardo, 69, 81
Villeda, Karen, 45, 93, 116
  Dodo, 21, 29, 95, 99–104
Vimeo, 137, 149, 161
violence, 1, 10, 50, 80, 197
  bare, 21, 58, 89, 175, 177
  Cuates de Australia (González) and, 128
  of dispossession, 137
  in *Dodo* (Villeda), 103
  domestic, 46

of extinction, 100
extractivist, 65, 89
gendered, 44
interpersonal, 101
against land defenders, 8, 21, 58–60, 62–64, 67, 82, 89, 175, 201n18, 201n24
language of, 87
in *Noche de fuego* (Huezo), 121
racialized, 191n14
rural resilience film and, 132, 144–46
slow, 21, 64, 89
state, 59–61, 86, 140
sustainable, 204

waste, 191n8, 197–98n74
  cinematic, 169
  electronic (e-waste), 159
  mine, 41, 69
  plastic, 38
  toxic, 3, 40, 49–51, 80
  workers, 160
whales, 96, 208n8
  gray, 91, 95, 209n20

YouTube, 149–50, 158, 161, 173

Zacatecas, 40–45, 80, 180, 194n48, 194nn52–53, 197n73
  FEMSA Biennial in, 53
  foreclosure of, 51
  See also Dávila, Amparo; mining; San Felipe Nuevo Mercurio
Zapata, Emiliano, 20, 58, 72
Zapata, Isabel, 21, 45, 92–93, 99–100, 115
  ballena es un país, Una, 208n8, 210n30
Zapatistas, 15–16, 204n56, 209n25
Zapotec, 1, 69, 167
  futurity, 6
  land defense and, 7, 81–82

Zavala, Oswaldo, 62, 64
Zelizer, Barbie, 19, 118
Zoque, 97, 98

www.ingramcontent.com/pod-product-compliance
Lightning Source LLC
Chambersburg PA
CBHW070757230426
43665CB00017B/2392